EXPERIMENTAL VACUUM SCIENCE AND TECHNOLOGY

AMERICAN VACUUM SOCIETY EDUCATION COMMITTEE

JOHN A. DILLON, JR., *Chairman*
University of Louisville
Louisville, Kentucky

WILLIAM F. BRUNNER
Lawrence Radiation Laboratory
Livermore, California

TEI IKI
Tektronix, Inc.
Beaverton, Oregon

JAHN M. KHAN
Lawrence Radiation Laboratory
Livermore, California

DAVID LICHTMAN
University of Wisconsin-Milwaukee
Milwaukee, Wisconsin

VIVIENNE J. HARWOOD, *Coordinator*
University of British Columbia
Vancouver, Canada

NORMAN MILLERON
Lawrence Radiation Laboratory
Berkeley, California

HUGH G. PATTON
Lawrence Radiation Laboratory
Livermore, California

M. TOM THOMAS
Washington State University
Pullman, Washington

NORMAN WILSON
University of New Mexico
Albuquerque, New Mexico

EXPERIMENTAL VACUUM SCIENCE AND TECHNOLOGY

EDITED BY

American Vacuum Society Education Committee

JOHN A. DILLON, JR., Chairman
VIVIENNE J. HARWOOD, Coordinator

MARCEL DEKKER, INC., New York   1973

COPYRIGHT © 1973 by MARCEL DEKKER, INC.

ALL RIGHTS RESERVED.

Neither this book nor any part may be reproduced or transmitted in any form or by any means, electronic or mechanical, including photocopying, microfilming, and recording, or by any information storage and retrieval system, without permission in writing from the publisher.

MARCEL DEKKER, INC.

95 Madison Avenue, New York, New York 10016

LIBRARY OF CONGRESS CATALOG CARD NUMBER: 72-97485

ISBN: 0-8247-6068-9

PRINTED IN THE UNITED STATES OF AMERICA

PREFACE

This book is intended as a source of ideas for those involved in the instructional aspect of vacuum science and technology rather than as a conventional laboratory manual. No branch of science is a closed book, and certainly vacuum science represents an area in which man has just begun to write. In the past two decades, we have made enormous progress in our ability to achieve rarefied atmospheres, and in the process, much of our naivete has also been pumped away. The ability to reduce the concentration of gas molecules in which we normally exist has established new degrees of freedom in many areas of study. At the same time it has often forcibly called attention to how very little we really know about the fundamental physical and chemical processes involved. For instance, it is now possible to produce ambients in which meaningful chemisorption experiments can be conducted, but our knowledge of the various forces which hold molecules to surfaces is still in its infancy.

The intrusion of vacuum methods and problems into a large number of disciplines has made it necessary to introduce an increasing number of science students to the basic concepts and techniques of the field. Frequently the suggestion has been made that a good collection of vacuum experiments would be most welcome, particularly if it could be drawn up so as to project some of the excitement of a lively field. With these ideas in mind, the Education Committee of the American Vacuum Society solicited the entire membership of the Society for suggestions concerning experiments which might prove valuable to those working with neophytes in the field. This book, therefore, represents the combined efforts of many people who are active workers in vacuum science and technology.

The Education Committee served as a board of editors in directing its activities not merely to the form of the contributions but to their substance as well. In some cases, suggestions for revision were made which increased the scope and the value of the experiments, and in other cases, members of the Committee appended editorial questions or comments of their own. The editors wanted to retain the original thoughts and words of the authors while at the same time introducing other points of view concerning the same material. It must be understood that such additions were intended not as criticisms or rebuttals, but as means of opening up new avenues of thought.

PREFACE

The Introduction attempts to set the tone for an unconventional excursion into a very complicated field. Section 1 presents some of the basic steps necessary in setting up the primary conditions necessary for performing more advanced experiments--the production and evaluation of a vacuum. In later sections, experiments are outlined which illustrate the characteristics of what is left behind, a few of the things which can be done in a vacuum, and some of the pitfalls which await the unwary. In each case an attempt is made to outline the background material and to suggest in rudimentary form how one might go about making the study. It is hoped and anticipated that the reader will devise alternatives to the suggestions made. Numerous references are also given which should assist both the student and his mentor in grasping what previously has been accomplished or at least attempted.

The reader will find the level of material presented to be quite nonuniform. Some of the experiments are very simple; some are extremely complicated. This was done deliberately, because the editors have attempted to produce a book which has appeal for a wide variety of users. It is not a training manual, a textbook, or a set of lecture notes. Rather it is a series of graded experiments from simple procedures to some sophisticated vacuum processes.

Since the book is not intended as a textbook itself, emphasis has been placed on providing as many pertinent references as possible. For instance, in the first experiment one will not find a diagram illustrating the internal operation of a rotary pump. However in the references cited, such diagrams and explanations abound. Rather than attempt an all-inclusive work, it was thought better to focus on the experiments while at the same time leading the reader into the literature of the field.

The members of the Education Committee are most appreciative of the contributions which the Society members have submitted and of the patience and forbearance they have shown. Special praise is due our colleague Vivienne Harwood who has served as co-ordinator, prodder, and provacateur in bringing the volume to completion.

Thanks are also due the American Institute of Physics, the Technical Publishing Company of Barrington, Illinois, and Johann Ambrosius Barth Publishers of Leipzig for permission to reprint certain materials.

Nobody knows the whole story in the vacuum field; that is the message which this book attempts to convey. Nevertheless, there are some interesting things which can be done, and we hope that the reader will share our enjoyment in attempting a few of them.

John A. Dillon, Jr.

INTRODUCTION

Since the basic material to which vacuum science addresses itself relates to rarified atmospheres, some of the fundamental ideas of the kinetic theory of gases are essential to a proper appreciation of what transpires inside any vacuum system. Because this book is intended for persons with a wide variety of backgrounds, some of the readers will be quite familiar with the concepts of kinetic theory while others will not. For the latter group, it is suggested that a simple presentation, such as that of Cowling,* or perhaps the type given in most introductory physics texts, be studied. The most essential ideas of the theory can be summarized as follows.

As in any physical theory, one starts with certain hypotheses which are framed in the light of experimental evidence. In this case the information comes from studies of the properties of gases and of chemical reactions which yield information concerning the ultimate constitution of matter. The principal hypotheses are these:

1. Gases are composed of atoms or molecules which are the smallest units of an element which retain the specific properties of that element.

2. These atoms or molecules are very small; evidence indicates dimensions of the order of angstrom units.

3. There are very many of these atoms per unit volume even at low pressures. This requirement arises from the fact that the properties of a gas are continuous throughout the gas.

4. On the average, there are large distances between these particles. The fact that gases are compressible demands this assumption.

5. The particles are in random motion and undergo elastic collisions with the walls of the chamber. Properties such as diffusion strongly suggest this internal motion.

---

*T.G. Cowling, <u>Molecules in Motion</u>, Harper, New York, 1960.

On the basis of these assumptions it is possible to show that

$$PV = 1/3 \, N \, \overline{mv^2} \qquad (1)$$

where P is the pressure and equals force/area at the walls; V is the volume of the gas; N is the number of atoms or molecules in the chamber; m is the mass per atom or molecule; and

$$\overline{v^2} = \frac{v_1^2 + v_2^2 + \cdots + v_N^2}{N}$$

where v_1, v_2, etc. are the speeds of the individual particles.

By combining this theoretically derived equation with the ideal gas law, it can be shown that:

$$1/2 \, \overline{mv^2} = 3/2 \, kT \qquad (2)$$

where k is Boltzman's constant (1.38×10^{-16} ergs/degree), and T is the absolute temperature in degrees Kelvin.

The importance of this relationship is that it identifies temperature as a measure of the internal energy of the atoms or molecules. Similarly the concept of pressure in a gas arises from the rate of change of momentum of the basic particles in their elastic collisions with the walls.

The overall picture is therefore a simple one. The atoms or molecules move about with velocities dependent on the temperature of the gas and on the masses of the particles involved, the larger masses moving on the average more slowly than the smaller masses in conformity with Eq. (2).

If one combines Eqs. (1) and (2), it can be seen that

$$PV = NkT \qquad (3)$$

From this equation it is seen that at a given temperature, the pressure is dependent upon the number of atoms or molecules per unit volume.

Even these simple aspects of the theory give a good indication of the basic physics which govern the properties of gases. However, as one examines the experimental evidence more closely, it becomes apparent that the assumptions we have made are indeed too simple. No account has been taken of the fact that the atoms or molecules themselves occupy space and therefore decrease the free volume which is available for motion. Nothing has been said of internal forces which might exist among the particles and therefore effect their motions. Both of these factors have to be accounted for at higher pressures, where the increased number of particles per unit volume makes them important. Also closer inspection of energy considerations

INTRODUCTION

indicates that modes such as rotation and vibration must be considered in addition to translational motion, which is all we have discussed. These necessary refinements can be made by improving upon our initial assumptions and in no way endanger the great value of the fundamental ideas of the theory. But, in addition to such improvements of the theory which have been made, there are other practical questions to which the vacuum worker must be ever alert. Are the walls of the vacuum chamber playing an active role in the system by serving as sources or sinks for gases? What do the chemical activities of the constituent gases do to processes, such as thermionic emission, which are being carried out in the chamber? In other words, are the components of a vacuum chamber, including the gases, idealized, inert units or do their activities sometime fall beyond the normal scope of the kinetic theory. Some of these questions are explored in the next section.

John A. Dillon, Jr.

A DIALOGUE ON VACUUM SCIENCE

Adopting a time-honored technique of presentation, here is a dialogue that might take place among an audience interested in vacuum technique, a panel consisting of two vacuum specialists, and a moderator.* Contributions by these imaginary persons are identified here by MOD, AUD, PAN I, and PAN II.

MOD: To promote free exchange on the general topic, "Vacuum," our panelists invite your questions and comments as well as theirs and mine.

AUD: What is a vacuum, anyway? I'm not satisfied with, "... a space devoid of matter" or "... a degree of gas rarefaction well below atmospheric pressure."

PAN 1: Perhaps vacuum like beauty, rests in the mind's eye of the beholder. One man's vacuum may be another's sewer, but the proof of understanding is the ability to PREDICT achievement, i.e., to know what needs to be done and how to do it in terms of time, manpower, cost, equipment, etc. In the practical world, here on earth, vacuum is achieved by pumping on a chamber.

Look at vacuum in terms of what is important to you. Do you care about the nature and number of molecules that are malingering on chamber walls, speeding about in space, and entering and leaving--in short, is it fruitful for you to view vacuum as a rarefied environment? The kinds of stuff and how much of each kind can be tolerated on the walls and in the gaseous phase are often found by trial and error.

PAN 11: Sounds at best like much ado about nothing. Why oppose the tradition of measuring vacuum in terms of the sum of the partial gas pressures?

PAN 1: If knowing the sum of the partial pressures were essential to predict when a vacuum would be satisfactory, there would be no problem. But if we take "gas pressure" seriously, we find we're getting onto the wrong track; i.e., we are focusing our attention on momentum

---

*This discussion has been reprinted from Research/Development 21, 40 (1970).

transfer rather than on more important parameters. In the most general sense, pressure in a gas is defined as "the time rate of transfer of momentum normal to a plane in a given direction."

AUD: I read in my physics book that "The pressure in a vacuum system is defined as the force exerted by the gas per unit surface area."

PAN 1: Your statement is an application of the more general definition just quoted above. Agreed?

PAN 11: Let's see what sort of trivial trap you are trying to bait. Substitute the weighted sum of the partial gas densities of concentrations for the sum of the partial pressures.

PAN 1: In order to predict what is likely to happen in a vacuum system, strive for something approximating an equation of state for the vacuum you need. Whenever necessary, take into account the adsorbed, absorbed, and gaseous phases as well as the probability of entrance and exit as a function of time, temperature, and molecular kind as well as amount. For example, turn to Redhead, Hobson, and Kornelsen, The Physical Basis of Ultrahigh Vacuum (Chapman and Hall, 1968). The whole of Part A can be cited in support of an environmental specification of vacuum. Focus attention upon this material and observe that in most applications the kinetic energies of motion of molecules are of less importance than are the energies of bonding and potential bonding among molecules. The importance of momentum transfer (pressure in a gas) is almost negligible. Among other properties of molecules that are of primary importance are their cross sections for interactions with electrons, ions, atoms, molecules, and photons as a function of mutual energy.

AUD: The American Vacuum Society (AVS) has issued a set of vacuum standards and a glossary of terms. Why hasn't an environmental approach been stressed by AVS?

PAN 1: Standards and customary practice are growing, not static. An environmental approach is perhaps due for further discussion.

AUD: AVS is affiliated with the American Institute of Physics. Won't labeling vacuum as an environmental discipline knock us vacuum types out of the scientific park into the engineering field?

PAN 1: Hardly a likely prospect. Just because vacuum is not a one-parameter show, doesn't mean that vacuum can't remain alongside the other multiparameter disciplines under the American Institute of Physics tent. Indeed, the environmental approach requires that attention be directed to the basic chemical and physical processes involved.

DIALOGUE

AUD: It appears then that the vacuum field is perhaps unique in modern technology: a colloquial nostrum, "gas pressure," is taken seriously as a candidate for the international vacuum standard.

PAN 1: You ought to be up here on this panel--Thanks.

PAN 11: Who says there's acceptance here of this nonsense about an environment? Fellow panelist, I challenge you to change the tradition that serves us all well . . .

PAN 1: No need to change habits of speech. One should avoid, however, erecting colloquialisms as marble standards.

AUD: A discouraging spectacle--our mentors talking past us at each other. Please get this discussion back on a practical level. What advantage, if any, does an environmental approach give over a traditional one?

AUD: Excuse me, please! Before I get lost, how about the traditional approach to vacuum?

PAN 11: Traditionally, vacuum is acknowledged to be created by pumping on a sealed chamber, the degree of vacuum increasing as the pressure exerted by the residual gas decreases below atmospheric. Measuring a system's absolute base pressure under no load conditions (clean, dry, and empty) is the traditional way to classify the degree of vacuum obtainable. For example, we speak of low, medium, high, and ultrahigh vacuum systems corresponding to regions of lower and lower base pressure capability. Under a gas load, the operating pressure is proportionally higher but the base pressure capability is taken as an index of the degree of control of contamination. Of course, residual gas analyzers can be used to indicate the partial pressures whenever necessary.

Much thought and effort over many years has resulted in a voluminous literature in firm support of the traditional approach to vacuum. Well founded custom must not be cast aside in favor of some crack pot philosophy that at best merely puts new labels on things already well established.

PAN 1: Well, I can agree that at least part of the vacuum literature is infirm. Results, not change should be sought. If a method-- whether new, old, or 'cracked'--serves in a given application, why not use it? And why not set aside a method wherever it is cumbersome or its limitations are manifest? Perhaps there is a straightforward explanation regarding why "pressure" methods have served and still do: many vacuum requirements are that crude. There's scarely another discipline where adverse changes of 10 or more in the index parameter can be tolerated. Let me offer that the traditional approach embraces the ancient (and still very useful) dogma, "Vacuum is an antifluid that flows and can be therefore piped and pumped." The garden variety of very handy vacuum formulas (which you can identify in numerous useful books) spring from this. For example, $Q = PS$, $Q = C(P_1 - P_2)$, $1/S_1 = 1/S_2 + 1/C$ are used with profit by all of us everyday. My word of caution is, recognize strengths and weaknesses in what you use.

AUD: Name some principal weaknesses.

PAN 1: In general, the gas concentration is nonuniform. Also, free molecular flow often isn't adequately described by the sum of the reciprocals approach. But permit me to try to explain by following the pump down of a chamber.

Start with a leaktight chamber filled with air. Identify the stuff clinging to the walls and in the gas phase as you pump down. Note that the longer time that vacuum pumps act on the room temperature chamber, the less the percentage composition of the chamber's gas phase resembles the composition of air. Indeed, by the time molecules in the gas phase begin to hit wall surfaces more often than they hit each other, about 80 to 90 per cent of the gas phase is water vapor. The water molecules come from the surfaces. As pumping is continued (almost invariably baking or cooling the chamber is required to minimize time), carbon monoxide is often most abundant. Further heating or cooling leaves hydrogen as the irreducible remnant in the gas phase--coming mostly from bulk materials and pumps, not from surfaces. The percentage composition of the surface phases rarely match the gas phase.

AUD: How can this behavior be predicted?

PAN 1: If you take an environmental approach, a qualitative prediction can be made by using pumping and outgassing data. But the realization of a quantitative prediction is much more difficult.

As free molecular flow conditions are achieved in the chamber, the predictions of kinetic theory correspond less to observation. The walls of the box are said to be outgassing. Indeed at a point in time the pump is operating not on the gas originally in the volume of the box, but rather on the gas coming from the adsorbed and absorbed phases associated with all surfaces and bulk materials inside the box. As time goes on the predictions of the kinetic theory differ from measured performance by many powers of ten. Indeed, if we ask the question, "When is the vacuum good enough to start a process or experiment?"--we must turn our back on "gas pressure" and focus our attention on far more important parameters of the environment.

AUD: Isn't it the vacuum pump that really creates the vacuum?

PAN 1: Well, whatever pump we use may ensure that the degree of vacuum desired is never achieved, i.e., all pumps eventually become sources of gas; and depending upon what environmental conditions

DIALOGUE   xiii

you need, molecules originating from the pumps themselves (some kinds are worse than others) can limit the environment.

AUD: How about the suction force created by pumps? Don't ultrahigh vacuum pumps suck harder on the walls of the box?

PAN 1: It's handy to use a thought like "suction" in colloquial speech, but let's realize that pumps cannot reach out and grab molecules. At present, pumps must wait for the molecules to reach their plane of action. As a footnote, it may be possible someday to reach out and grab molecules in the vacuum space, but don't hold your breath waiting.

AUD: But doesn't the gas pressure gradient drive the molecules into the pump?

PAN 1: No. Not after free molecular conditions prevail. The temperature of the walls is then alone responsible for gas motion. Under free molecular conditions, the gas molecules move about just the same whether or not the pumps are on.

AUD: How about the conductance of the pipe connecting a chamber to a pump? I've heard that pumps can't do any better than the pipe used to hook them up.

PAN 1: Let's consider the plumbing which connects a pump and a chamber to include traps, baffles, valves, etc. One of the myths of current practice is that the temperature of this plumbing between the pump and a chamber affects the rate of free molecular flow in steady state. For engineering purposes the wall temperature of the plumbing can at most affect the flow by 10 per cent or less, provided that the temperature of the plumbing is not low enough to condense gas (producing a pumping effect) or high enough so that the plumbing breaks up molecules, reacts chemically, or outgasses at a high rate.

To get some idea of the conductance or impedance of plumbing, consider steady-state free molecular flow through a right circular cylinder of length equal to its diameter. Such a piece of pipe has the appearance of a napkin ring. If we apply our laminar flow intuition, such a pipe would seem to offer little or no impedance to the flow of free molecular gas. This intuitive sense of laminar flow draws us into grave error, for if such a piece of pipe connects two large volumes, each under free molecular flow conditions, about one-half of the molecules entering such a pipe in steady state will be rejected by wall interactions and will be returned to the entrance through which they started. This means that no matter how large a pump is placed at the end of a pipe having a length equal to the pipe diameter, only about one-half of the flow entering the pipe can be pumped away.

Clearly, the impedance of the plumbing is of overriding importance to the system speed. As a handle for the intuition, typical vacuum pumping systems are less than 10 per cent efficient; i.e., less than 10 per cent of the molecules entering the hole in the side of the chamber wall leading to the pumping system are pumped. In other fields of technology such poor performance would not be tolerated. This sort of poor performance is tolerated in the vacuum business largely because at present, pumping efficiency is not expressed in terms of what is going on in the chamber, but in terms of what is going on at some point of little or no consequence to the user of the chamber.

AUD: What about expressing pumping speed in volumetric units?

PAN 1: In the general case in which non-Maxwellian gases are reaching a pump, volumetric units won't do. Pumping speed can then be measured unambiguously in terms of the number of molecules pumped per unit time. In the specific case in which a sufficiently Maxwellian gas is reaching a pump, volumetric units serve nicely.

AUD: What about gauges and manometers?

PAN 11: All gauges are manometers.

AUD: Ionization gauges seem to serve well enough. How is this possible if they don't indicate gas pressure?

PAN 1: Fortunately, in many work-a-day situations hot-filament ionization manometers provide a crude assessment of vacuum as an environmental condition--a fact that is much more important than their lack of accuracy in measuring the pressure or density. Matters would be much worse if gadgets reading true momentum transfer were the only available vacuum gauges. However, be alert to the fact that gauges using electron bombardment ionization, whether of the hot- or cold-filament variety, can alter the environment they are directed to measure.

AUD: What about the vapor pressure of components in my vacuum?

PAN 11: Beware of "vapor pressure." Most of the time this term is misapplied; for example, people often say, "The vapor pressure of plastic (or rubber, or metal, or some other substance) is too high." Think about the definition of equilibrium vapor pressure and you'll see that irreversible decomposition and contaminent outgassing are more appropriate terms than vapor pressure.

AUD: What about the outgassing of stainless steel, aluminum, iron, etc?

PAN 1: This outgassing depends far more on the _history_ of the material than on some irreducible property.

AUD: You mean all the manufacturing steps and the finger prints?

DIALOGUE

PAN 1: Yes.

AUD: What about clean surfaces? Aren't clean surfaces required in vacuum?

PAN 1: If you use the definition of clean that surface science uses, no. Surface scientists define an atomically clean surface as made up of molecules of the same kind as in the bulk material. Vacuum people, however, often want a nonactive surface that is stable under photon, electron, and other energy bombardment. Common sense indicates, however, that sewage isn't tolerable even in amounts you can't see with your eye or pressure gauge. Don't take the clean, dry, and empty convention too seriously.

MOD: Regretfully, I see our time is up. I hope that enough loose ends are left flying to stimulate each of us to further discussion. Let's not think in terms of a rigid approach to vacuum, but search for how to predict what is likely to happen in one's own vacuum environment. Permit me to offer my own "open end": vacuum is a molecular environment sufficiently more rarefied than are its surroundings.

                                                       Norman Milleron

CONTRIBUTORS TO THIS VOLUME

G. BREITWEISER, Sloan Instruments Corporation, Santa Barbara, California

W. F. BRUNNER, Lawrence Radiation Laboratory, Livermore, California

K. M. BUSEN, Williams College, Williamstown, Massachusetts, and Sprague Electric Company, North Adams, Massachusetts

M. R. CARBONE, Mason-Renshaw Industries, Carpinteria, California

H. FARBER, Department of Electrophysics, Polytechnic Institute of Brooklyn, Graduate Center, Farmingdale, New York

P. GROSEWALD, IBM Watson Research Center, Yorktown Heights, New York

B. R. F. KENDALL, Department of Physics and Ionosphere, Research Laboratory, Pennsylvania State University, University Park, Pennsylvania

J. G. KING, Massachusetts Institute of Technology, Cambridge, Massachusetts

R. P. W. LAWSON, Electrical Engineering Department, University of Alberta, Edmonton, Canada

H. M. LUTHER, Department of Physics and Ionosphere, Research Laboratory, Pennsylvania State University, University Park, Pennsylvania

P. E. McELLIGOTT, General Electric Company, Schenectady, New York

J. R. MILLER, III, U. S. Army Metrology and Calibration Center, Redstone Arsenal, Alabama

N. MILLERON, Lawrence Radiation Laboratory, Berkeley, California

C. F. MORRISON, Granville-Phillips Company, Boulder, Colorado

R. OLSON, Department of Physics, San Fernando Valley State College, Northridge, California

J. ORSULA, Massachusetts Institute of Technology, Cambridge, Massachusetts

Wm. N. PARKER, RCA Electronic Components, Lancaster, Pennsylvania

H. G. PATTON, Lawrence Radiation Laboratory, Livermore, California

R. P. RIEGERT, Sloan Instruments Corporation, Santa Barbara, California

F. ROSEBURY, Research Laboratory of Electronics, Massachusetts Institute of Technology, Cambridge, Massachusetts

M. T. THOMAS, Department of Physics, Washington State University, Pullman, Washington

K. B. WEAR, 1114 Fayetteville Rd. S.E., Atlanta, Georgia

G. K. WEHNER, Department of Electrical Engineering, University of Minnesota, Minneapolis, Minnesota

D. WHITCOMB, Motorola, Inc., Phoenix, Arizona

J. WILSON, Department of Physics, San Fernando Valley State College, Northridge, California

CONTENTS

Preface	iii
Introduction	v
Dialogue	vii
Contributors	xvii

Section 1

PROCEDURES IN VACUUM PRODUCTION AND MEASUREMENT	1
Experiment 1.1 Rotary Oil-Sealed Mechanical Vacuum Pump	3
W. F. Brunner and H. G. Patton	
Experiment 1.2 Oil Vapor Diffusion Pump	17
W. F. Brunner and H. G. Patton	
Experiment 1.3 Cryosorption Pump	25
W. F. Brunner and H. G. Patton	
Experiment 1.4 Getter-Ion Pump	33
W. F. Brunner and H. G. Patton	
Experiment 1.5 Vacuum Measurement Techniques	39
W. F. Brunner and H. G. Patton	

Section 2

CHARACTERISTICS OF THE VACUUM ENVIRONMENT — 60

Experiment 2.1 Demonstration of the Outgassing of Different Vacuum Materials — 61

F. Rosebury

Experiment 2.2 Comparison of Gas Evolution Phenomena from Glass and Metal Vacuum System Envelopes during Baking — 65

R. P. W. Lawson

Experiment 2.3 Determination of the Net Quantity of Gas Flowing through a Cylindrical Tube — 71

K. M. Busen

Section 3

STUDIES OF THE DEPENDENCE OF PHYSICAL PROPERTIES OF GASES ON GAS DENSITY — 82

Experiment 3.1 Measurement of the Pumping Action of an Ionization Gauge — 83

H. Farber

Experiment 3.2 Study of the Linearity of an Ionization Gauge — 89

J. R. Miller, III

Experiment 3.3 Calibration of Gauges — 93

C. F. Morrison

Section 4

PHYSICAL AND CHEMICAL INTERACTIONS AT SURFACES — 108

Experiment 4.1 Study of the Sorption of Gases for Different Gas-Sorbent Combinations — 109

K. B. Wear

CONTENTS

Experiment 4.2	The Use of Sorbents as Traps and Pumps	115
	H. Farber	
Experiment 4.3	Sorption of Gases by Titanium	123
	H. Farber	
Experiment 4.4	Investigation of the Passage of Oxygen across a Silver Barrier	133
	K. M. Busen	

Section 5

PROCESSES REQUIRING A VACUUM ENVIRONMENT — 142

Experiment 5.1	Thin Film Evaporation	143
	M. T. Thomas	
Experiment 5.2	Fabrication of a Nichrome Resistor	149
	R. P. Riegert and G. Breitweiser	
Experiment 5.3	Sputtering	153
	P. Grosewald	
Experiment 5.4	Ejection Patterns in Single Crystal Sputtering	161
	G. K. Wehner	

Section 6

SPECIAL PROJECTS — 164

Experiment 6.1	Study of the Sublimation of Ice at Various Pressures	165
	Wm. N. Parker	

Experiment 6.2	Study of Friction	171
	P. E. McElligott	
Experiment 6.3	Measurement of the Mean Free Path of Conduction Electrons in Silver	181
	R. Olson and J. Wilson	
Experiment 6.4	Construction and Use of a Cathode Ray Tube	193
	B. R. F. Kendall and H. M. Luther	
Experiment 6.5	Construction of a Vacuum Triode Using Glass Techniques	201
	J. G. King and J. Orsula	
Experiment 6.6	Experiments Using Solder Glass Techniques	215
	D. Whitcomb	

Section 7

SPECULATIONS 220

Experiment 7.1	Original Thought Experiments	221
	M. R. Carbone	
Experiment 7.2	Provocative Ideas and Questions	225
	N. Milleron	

BIBLIOGRAPHY 227

APPENDIX 231

Index 235

Section 1

PROCEDURES IN VACUUM PRODUCTION AND MEASUREMENT

Experiment 1.1

ROTARY OIL-SEALED MECHANICAL VACUUM PUMP

W. F. Brunner and H. G. Patton

Lawrence Radiation Laboratory
Livermore, California

INTRODUCTION

This experiment describes the operating principles and procedures for the mechanical pump, the proper application of gauges suitable for the pressure range in which the pump normally operates, and the limits imposed on the effectiveness of the pump by connecting tubing.

The selection of a small pump (8 cfm or less) will simplify the performance of this experiment. The manufacturer's instruction manual should be used to establish pumping capacity and other general operating characteristics.

OPERATING PRINCIPLES

A basic element in almost all vacuum laboratories, the oil-sealed rotary vacuum pump is a positive displacement pump capable of having an extremely high compression ratio. It will discharge against atmospheric pressure and can produce a total pressure of a few hundreths of a Torr or less. Its operating principle is as follows: Gas from the vacuum chamber enters the pump through the inlet port. The gas is trapped by the rotor-stator arrangement, compressed, and swept toward the discharge port. When the pressure of the trapped gas is raised slightly above one atmosphere, the discharge valve will lift and the gas will be expelled to the atmosphere.

EXPERIMENTAL SYSTEM

Assemble a small manifold consisting of a Bourdon gauge, a vacuum valve, and some connecting fittings, and attach this assembly to the inlet of the mechanical pump as shown in Fig. 1. Do not allow anything to fall into the inlet of the pump. Before operating, inspect the pump for: broken or loose V-belt, incorrect oil level, loose motor mounts, and excessive oil leakage.

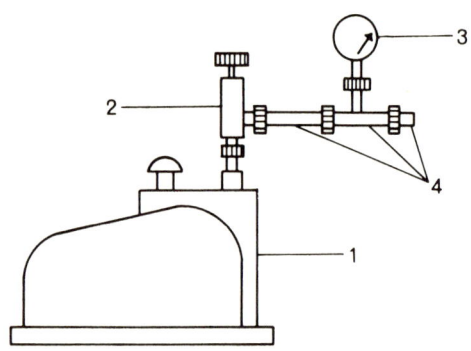

FIG. 1. Mechanical pumping system. Equipment required: 1. mechanical pump, 2. mechanical pump valve, 3. Bourdon gauge, 4. connecting fittings.

Possible Hazards

    Exposed pulleys and V-belts
        (should be covered by a guard)

    Oil ejected from the exhaust port
        (exhaust baffle should be in place)
        (oil level should not be too high)

When the power switch is activated the Bourdon gauge should rapidly indicate a low pressure. If not, check for leaks in the gauge manifold, incorrect oil level, or improper pump rotation.

1.1. ROTARY OIL-SEALED MECHANICAL VACUUM PUMP

EXPERIMENTAL PROCEDURES

A. EXAMINATION OF PUMP PERFORMANCE

Assuming the pump does produce a very low vacuum on the Bourdon gauge, it can be presumed that the pump is operating correctly. However, correct operation does not mean the performance will be in accordance with the manufacturer's published specifications. The two operational characteristics of mechanical pumps that most closely define performance are the ultimate pressure and the volumetric pumping speed. Detailed procedures for defining the performance characteristics of mechanical pumps are covered by the AVS Standards-- T.S. 5.1 and T.S. 5.2. Using either of these sources it is possible to closely define the performance of a pump. However, the following less exact procedures will provide a close approximation.

The ultimate pressure quoted by the manufacturer usually refers to pressure measured with a McLeod gauge. When McLeod gauges are used for this purpose, some sorption and compression of vapors occur during measurement. Therefore, ultimate pressures obtained will be lower than those obtained with a thermocouple or Pirani gauge. The inclusion of a trap between the pump and gauge will give more consistent information about the ultimate pressure to be expected from a well-trapped mechanical pump. By including a U-trap in the connecting tubing between the pump and chamber, both untrapped and trapped ultimate pressure can be measured. First, the untrapped measurement is obtained, then the U-trap is cooled with a suitable refrigerant (ice water, dry ice in acetone slush, or liquid nitrogen) and the trapped measurement is obtained.

Assemble the equipment as shown in Fig. 2, using tubing as large as, or larger than, the diameter of the inlet to the pump. The connecting lines should be kept short to minimize conductance losses.

Turn on the pump and evacuate the manifold. After the pressure has stabilized, the untrapped ultimate pressure has been attained. Record it and make a comparison with the manufacturer's published information on untrapped ultimate pressure for the pump being used

Now fill the refrigerant container and immerse the U-trap in it. This should result in a decrease in the pressure. Wait until the pressure stabilizes at the new level, record it as the trapped ultimate pressure, and compare it with the manufacturer's information.

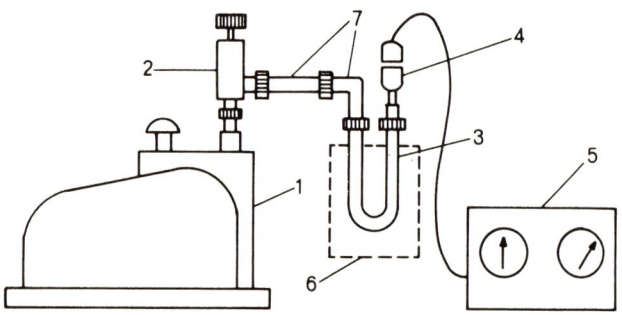

FIG. 2. Mechanical pump and trap. Equipment required:
1. mechanical pump, 2. mechanical pump valve, 3. U-trap, 4. thermal conductivity gauge tube, 5. thermal conductivity gauge control, 6. refrigerant container.

QUESTIONS

Suppose the stabilized untrapped pressure measured by this procedure is higher than the untrapped ultimate pressure quoted by the manufacturer. What are some of the conditions that could produce this situation?

What methods could you use to determine whether or not a higher than normal pressure was due to a leak?

Suppose there was a possibility that the pump had been exposed to a substance of high vapor pressure. How could you find out whether or not this substance was responsible for a higher than normal base pressure?

Should the stabilized trapped pressure be higher than the trapped ultimate pressure quoted by the manufacturer? What do you feel might be wrong?

B. MEASUREMENT OF PUMPING SPEED

The pumping speed listed by the manufacturer for any given type

1.1. ROTARY OIL-SEALED MECHANICAL VACUUM PUMP

of pump is usually the free air displacement at STP.* As the pressure decreases from atmospheric, there will be a reduction in the amount of gas pumped per unit of time (the mass flow rate). However, the pumping speed (volumetric flow rate) decreases only slightly until a pressure of around one Torr. The decrease then usually becomes more rapid, depending on the type of pump, and falls to zero at the ultimate pressure.

Determining the speed of a pump can be accomplished by using either a constant volume or a constant pressure method. The constant volume method is most frequently used in the range between atmospheric pressure and one Torr. Essentially, this method consists of measuring the time required for the pump to reduce the pressure a certain amount. The speed, S_p, is then calculated from the equation:

$$S_p = 2.3 \frac{V}{t_2 - t_1} \log_{10} \frac{P_1}{P_2}$$

where V is the volume of the enclosure, t_1 is the time at pressure P_1, and t_2 is the time at pressure P_2.

The constant pressure method is the most frequently used over the range between one Torr and the ultimate pressure of the pump. To determine the speed of a pump by the constant pressure method, a measured amount of gas Q is admitted to the enclosure to establish the pressure P. The speed, S_p, is then obtained from the equation:

$$S_p = \frac{Q}{P}$$

Possible Hazards

The U-tube and Dubrovin gauges use mercury as a fluid. Mercury is a cumulative poison. It is sufficiently toxic that knowledgeable help should be sought in cleaning up spills. Care should be taken when venting the chamber and gauges to atmospheric pressure in this exercise. Vent the gauge slowly or violent action of the mercury may break the gauge tubes.

---

*Standard temperature and pressure: $0°$ Centigrade and 760 Torr.

1. **Pumping Speed by the Constant Volume Method**

   Assemble the equipment as shown in Fig. 3, making the connecting tubing as short as possible to minimize line impedance. (Remember that any leaks in the assembly may lead to inaccuracy in the data to be collected.)

   Start this procedure with the mechanical pump operating and up to temperature, and the mechanical pump valve closed. Measure the time required to reach 100 Torr when the mechanical pump valve is opened. Obtain consistent readings for several consecutive runs. Now measure the time required to reach 10 Torr from 100 Torr. Again obtain consistent readings for several consecutive runs. Repeat the procedure for the time from 10 Torr to 1 Torr.

   From this information calculate the speed for each of the increments above, using the constant volume equation:

   $$S = 2.3 \frac{V}{t_2 - t_1} \log_{10} \frac{P_1}{P_2}$$

   Plot the calculated speed against pressure using semilog paper. (It is possible to make the curve more representative by using shorter increments and thus provide a greater number of points to plot.) In plotting, remember that the speed obtained by the calculation is the average speed over the pressure range from P_1 to P_2, thus the pressure used to plot the curve should be the average of P_1 and P_2.

   Replace the vacuum chamber used above with a smaller or larger chamber and repeat the procedure just completed. Plot the data and compare the two curves.

   QUESTIONS

   How do the speeds you have measured compare to those listed by the manufacturer over this pressure range?

   If there is a significant variation, can you provide an explanation?

   Is there any difference between speed calculated on the different chambers?

   Would you expect them to be the same or different? Why?

1.1. ROTARY OIL-SEALED MECHANICAL VACUUM PUMP

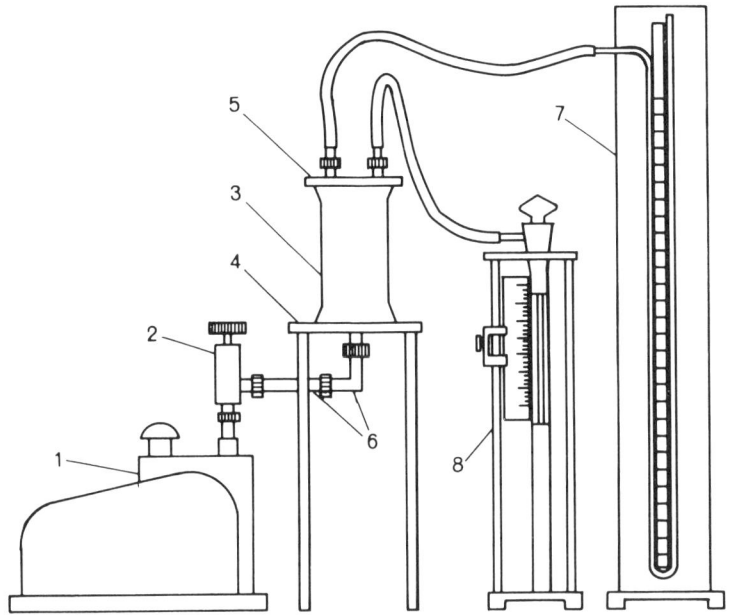

FIG. 3. Vacuum system for constant volume method. Equipment required: 1. mechanical pump, 2. mechanical pump valve, 3. vacuum chamber, 4. base plate and support stand, 5. cover plate and feedthroughs, 6. connecting fittings. In addition, there must be suitable gauges to measure from atmospheric pressure to one Torr, e.g., a U-tube manometer or a Dubrovin gauge (7 and 8).

2. Pumping Speed by the Constant Pressure Method

Assemble the equipment as shown in Fig. 4. The connecting tubing must be kept short as before to provide maximum conductance. The leak valve should be closed, the atmosphere valve open. The fluid level in the flowmeter reservoir should be high enough to cover the lower end of the tube when the fluid is at maximum height in the tube during a measurement.

Start the mechanical pump and open the mechanical pump valve.

Wait for the pressure to reach a stable low level before beginning the required measurements. It will take longer for the pressure to reach this level than it did during the procedure for obtaining the ultimate pressure of the pump. This is to be expected because of the increased gas load created by the outgassing of the chamber surfaces.

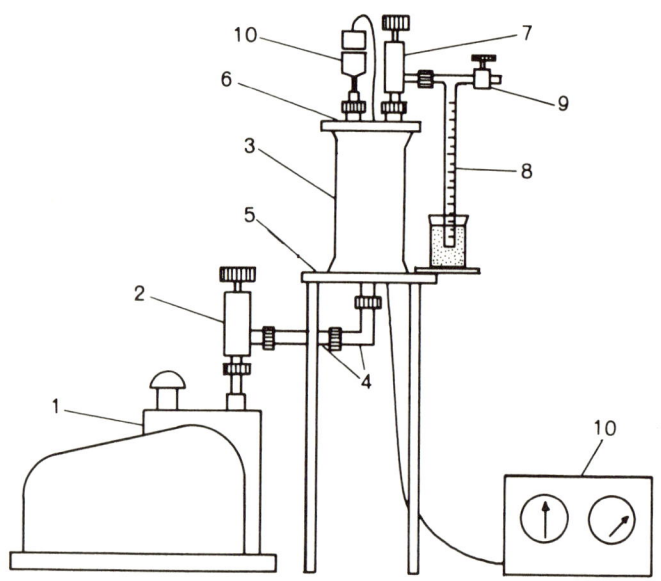

FIG. 4. Vacuum system for constant pressure method. Equipment required: 1. mechanical pump, 2. mechanical pump valve, 3. vacuum chamber, 4. connecting fittings, 5. base plate and support stand, 6. cover plate and feedthroughs, 7. leak valve, 8. flowmeter, 9. atmosphere valve. There must also be a suitable vacuum gauge to measure pressure between 1 Torr and 1×10^{-3} Torr, e.g., a thermocouple gauge or a Pirani gauge (10).

When the level is established, and with the atmosphere valve open, the leak valve is opened slowly and adjusted to maintain the pressure in the chamber at the level desired for the first speed test. Several minutes should be allowed for the conditions to stabilize. Now close the atmosphere valve and the air entering the chamber will be drawn from the flowmeter. The fluid will rise in the

1.1. ROTARY OIL-SEALED MECHANICAL VACUUM PUMP

buret to replace it. Measure the time required for the level to rise a specified amount. This volume of fluid represents the amount of air that leaked into the chamber during the measured interval of time. Thus, since the pressure and volume of the entering air are known, as well as the pressure in the enclosure, the rate at which the pump is removing air can be calculated from the equation:

$$S = \frac{Q}{P}$$

where

$$Q = \frac{\text{atmospheric volume } (V_a) \times \text{atmospheric pressure } (P_a)}{\text{time in seconds } (t) \text{ for } V_a \text{ to enter the chamber}}$$

and   P is the pressure in the chamber during the test.

Repeat this procedure at various pressure levels (at least four) between 10^{-3} and 1 Torr.

Plot these calculated speeds against pressure using semilog paper.

Replace the vacuum chamber used above with a smaller or larger chamber and repeat the procedure just completed. Plot the data and compare the two curves.

QUESTIONS

How do the speeds you have measured compare to those listed by the manufacturer over this pressure range?

If there is a significant difference between them, can you provide an explanation?

Is there any difference between speeds calculated on the different size chambers?

What reason could you give for a difference?

How do the speeds obtained in this procedure between 1 Torr and 1×10^{-3} Torr compare with those obtained in the last procedure, which were between 760 Torr and 1 Torr?

C. MEASURING THE EFFECT OF CONNECTING LINES ON PUMPING SPEED

When a section of line is inserted between a pump and a chamber it is evacuating, the time to pump this chamber down from some pressure P_1 to a lower pressure P_2 is increased. Obviously the impedance introduced by the added line has decreased the effectiveness of the pump at the chamber. Often the effect of the impedance can be predicted from the equation of the vacuum system:

$$\frac{1}{S_t} = \frac{1}{S_p} + \frac{1}{C_t}$$

However, an appreciation for the actual effects as measured rather than calculated can be obtained from the procedures outlined in the following sections.

1. During Viscous Flow

Reassemble the equipment as shown in Fig. 3, but add a line with a 1/2-in. inside diameter, 24 in. long, in series with the existing tubing connecting the pump to the chamber.

Following the procedure given with Fig. 3 (Pumping speed by the constant volume method), measure the time required to cover the same pressure increments. Obtain consistent readings for several consecutive runs over each of the increments to improve accuracy, and calculate the speeds.

Compare the speeds obtained above with the speeds obtained without the added line.

Replace the 1/2-inch by 24-in. line with a line 1/8 in. by 24 in. and perform the same tests.

Replace the 1/8-in. by 24-in. line with one that is 1/8 in. by 12 in. and obtain speeds over the pressure increments.

Plot the speeds calculated from these last three runs with the speed calculated from the pump alone on semilog paper.

Replace the pump used above with a smaller or larger pump and repeat the above procedure; first find the speed of the pump alone, then find the speed of the pump impeded by three different size lines.

1.1. ROTARY OIL-SEALED MECHANICAL VACUUM PUMP

QUESTIONS

The line diameter was reduced by a ratio of four to one. What is the ratio of the reduction in pump speed?

The line length was reduced by a ratio of two to one. What is the ratio of the reduction in pump speed?

Having noted the conditions produced by a change in pump size with a given line, comment on the predictability of changing the pump size to reduce the pump down-time.

The effect of bends in the connecting line can be explored by measuring the pump down-time through a coiled line of the same length and inside diameter as any of the straight lines used on the previous procedures.

2. During Molecular Flow

Assemble the equipment as shown in Fig. 5, using a 1/2-in. by 24-in. long line connecting the pump and the chamber. Use two gauges for this procedure located as shown in Fig. 5.

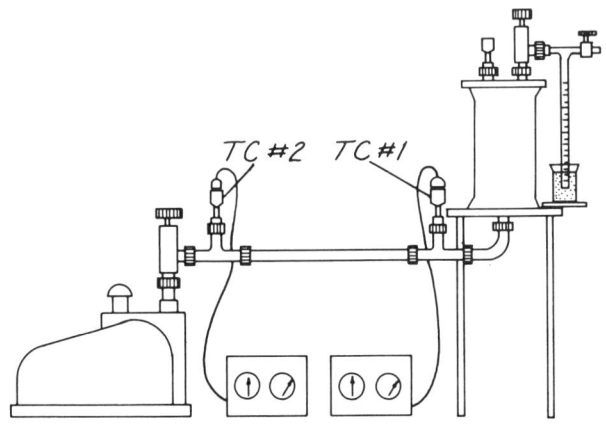

FIG. 5. Molecular flow system using two thermal conductivity gauges.

Using a leak valve and the techniques used in the section on pumping speed by the constant pressure method, establish a steady state pressure P_1 measured by gauge 1, the lowest pressure at which a leak rate can be measured. Measure the flow rate Q with the buret as before. Measure the pressure P_2 at gauge 2. Using pressures P_1 and P_2, calculate the speeds S_1 and S_2, using the formula:

$$S = \frac{P}{Q}$$

Calculate the conductance of the 1/2-in. by 24 in. line in three different ways:

$$C' = \frac{Q}{P_1 - P_2}$$

$$C'' = \frac{S_2 \times S_1}{S_2 - S_1}$$

$$C = C_m + C_v$$

(C_m is molecular conductance, C_v is viscous conductance)

Repeat the procedure above used with the 1/2-in. by 24-in. line using a small diameter, 24-in. long tube (1/8-in. to 3/16-in. metal tube or glass pipe).

Shorten the small diameter line to 12 in. and repeat the procedure.

QUESTIONS

Do the measurements you have made substantiate the relationship of conductance to length and diameter expressed by the standard equations?

Determine the effect on pumping speed and conductance of a line with several bends.

For the pump you have been using and the 24-in. line, what diameter would be necessary to maintain at least 80% of the pump speed at the chamber end of the line? (20% pumping speed loss is generally considered good design criteria.)

1.1. ROTARY OIL-SEALED MECHANICAL VACUUM PUMP

For 1/8-in. by 24-in. line, is there a point at which an increase in pump capacity will not yield a significant increase in total speed?

BIBLIOGRAPHY

1. W.F. Brunner, Jr., and T.H. Batzer, Practical Vacuum Techniques, Reinhold, New York, 1965, pp. 26-35.

2. B.D. Power, High Vacuum Pumping Equipment, Chapman and Hall, London, 1966, pp. 1-43.

3. C.M. Van Atta, Vacuum Science and Engineering, McGraw-Hill, New York, 1965, pp. 283-285.

Experiment 1.2

OIL VAPOR DIFFUSION PUMP

W. F. Brunner and H. G. Patton

Lawrence Radiation Laboratory
Livermore, California

INTRODUCTION

The AVS defines a diffusion pump as: "A vapor pump having a boiler pressure of the order of a few Torr and capable of pumping gas with full efficiency at intake pressures not exceeding about 20 millitorr and discharge pressures not exceeding about 500 millitorr."

This experiment describes operating principles and procedures for the diffusion pump, the proper application of gauges suitable for the pressure range in which the pump normally operates, and the limits imposed on the effectiveness of the pump by the addition of impedances in the form of baffles, valves, traps, and connecting tubing.

The vapor diffusion pump is the vacuum pump most frequently chosen when it is necessary to obtain useful pumping speeds at pressures less than 10^{-3} Torr. These pumps are quite simple in operation, the pumping action is provided by directed streams of fast-moving vapor molecules which strike gas molecules and give them momentum in the direction of the stream. With appropriate design this arrangement has been made to produce an extremely high compression ratio. However, the vapor diffusion pump is only able to compress the gas to a maximum pressure of about 5×10^{-1} Torr. It must therefore be supported, that is, "backed" by another pump that can exhaust to atmospheric pressure. The oil-sealed mechanical pump of Experiment 1.1 is most frequently used for this purpose.

OPERATING PRINCIPLES

The operating principle of the diffusion pump is as follows: Gas molecules diffusing into the inlet port are struck by molecules of pump fluid streaming from the jets of the pump. This gives the gas molecules momentum in the direction of the pump outlet, or discharge port. The action of several successive jet stages increases the pressure in the region of the pump outlet. This gas is then removed by the "backing pump." The pumping action of the diffusion pump is provided by molecules of pump fluid issuing from the jets of the pump as a directed stream of vapor. This vapor is supplied by heating a fluid in an integral boiler located in the bottom of the pump. The fluid is heated in the boiler to a temperature at which its evaporation rate will provide an adequate amount of vapor to supply the jets. The high velocity stream of vapor molecules leave the pump jets traveling outward and downward, striking gas molecules in their path, and finally condensing on the cool pump wall. After condensation, the fluid flows down the wall into the boiler to be reheated for another pumping cycle. The fluid used in a vapor diffusion pump may be mercury or one of a wide range of liquids generally and loosely classified as "oil." The choice of which fluid to use is most appropriately determined by the intended application.

EXPERIMENTAL SYSTEM

The set-up for this experiment will be simplest if a small, air-cooled diffusion pump can be obtained. Hopefully, it will also be possible to obtain a manufacturer's instruction manual from which the pumping capacity and other operating characteristics can be established.

Mount the diffusion pump on a test stand as shown in Fig. 1. Connect the mechanical pump to the forearm of the diffusion pump with a section of vacuum line. A mechanical pump valve and thermal conductivity gauge should be included in this line. Cover the inlet of the diffusion pump with a Pyrex plate and evacuate this assembly to about 3×10^{-2} Torr.

Possible Hazards

The exposed surface of the diffusion pump in the region of the boiler normally runs quite hot (200°C). It is advisable to provide a guard on the stand to prevent accidental contact. It is also advisable to have the test stand high enough to avoid setting fire to a bench or counter top.

1.2. OIL VAPOR DIFFUSION PUMP

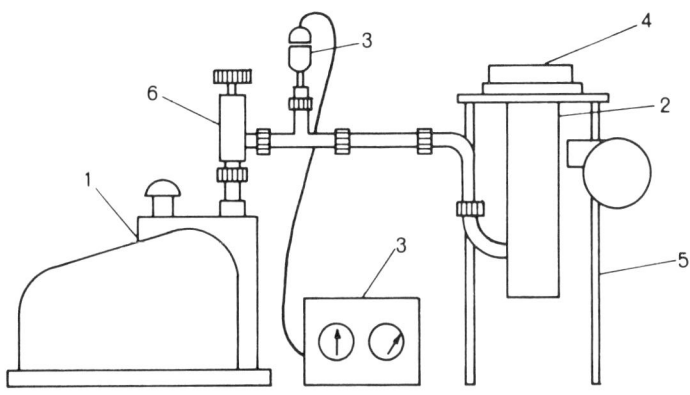

FIG. 1. Diffusion pumped system. Equipment required: 1. mechanical pump, 2. oil vapor diffusion pump, 3. thermal conductivity gauge, 4. Pyrex plate, 5. test stand, 6. small vacuum valve.

EXPERIMENTAL PROCEDURES

A. EXAMINATION OF PUMP PERFORMANCE

When the diffusion pump heater is turned on, observe the action within the pump as the fluid heats up. Record your observations and indicate what you feel that action is caused by. Note the change in condition of the Pyrex window as the pump heats. For example, how much time does it take for a slight haze to form, for a heavy haze to form, for small drops to form, for large drops to form: This condition represents the backstreaming that could be expected from this pump during start-up. If the condensed oil film is now evaporated from the inside surface of the port with a heating device such as a heat lamp, heat gun, or heating pad, it will be possible to see a fully operating diffusion pump in action.

Now cool the diffusion pump, vent it to atmospheric pressure, and change the test set-up to that shown in Fig. 2.

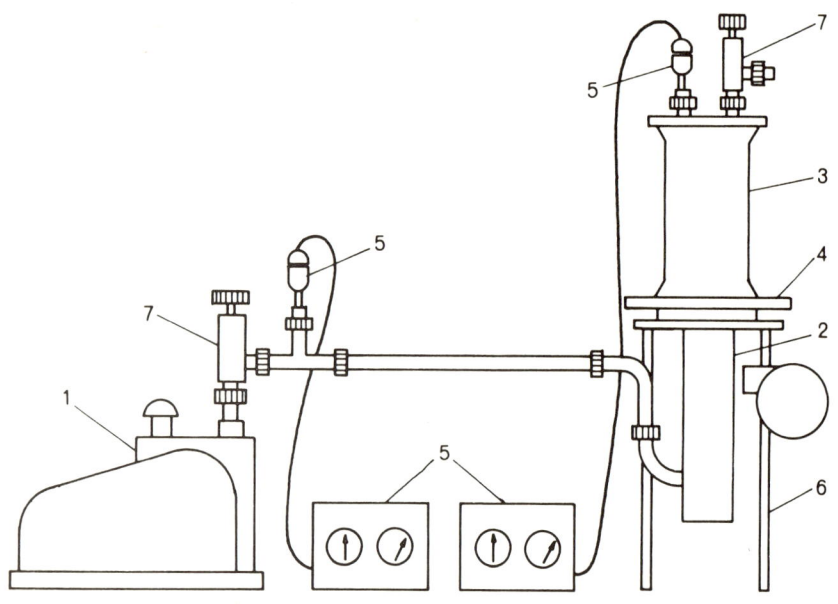

FIG. 2. Diffusion pumped system with vacuum chamber. Equipment required: 1. mechanical pump, 2. oil vapor diffusion pump, 3. vacuum chamber, 4. base plate, 5. thermal conductivity gauge, 6. test stand, 7. small vacuum valve.

Remove the Pyrex port from the diffusion pump inlet and replace it with a base plate. Install a vacuum chamber on the base plate. Place a cover plate fitted with a vent valve and thermal conductivity gauge tube on top of the vacuum chamber. Start the mechanical pump and reduce the pressure in the vacuum chamber (P_1) to 3×10^{-2} Torr. Note the pressure at the mechanical pump (P_2) at this time. Turn on the diffusion pump and fill out a table of pressure versus time as the pump becomes operational. Note any change in conditions that appear significant. Be particularly observant of coincident pressure changes in vacuum chamber and foreline. What conditions could be responsible for these changes?

1.2. OIL VAPOR DIFFUSION PUMP

Possible Hazards

Pyrex pipe is exceptionally tough and does not constitute an implosion risk: however, it can be chipped if handled carelessly. This alters the strength unpredictably as well as creating painfully sharp edges.

The test stand should provide the stability necessary to keep the assembly from becoming top-heavy.

QUESTIONS

After the pump was turned on, how long did it take before the pump began pumping?

What particular condition caused you to believe the pump began operating at this time?

Was the onset of pumping sudden, or did it drag out for some period of time?

B. MEASUREMENT OF PUMPING SPEED OF PUMP

Allow the pump to cool again and vent it to atmosphere. Install a leak valve and ion gauge in the top plate of the vacuum chamber. Re-evacuate the vacuum system and start the diffusion pump. Using the constant pressure, metered leak technique described in Experiment 1.1, measure the pumping speed of the pump over the range 1×10^{-5} Torr to 5×10^{-4} Torr. Use at least four pressure levels and plot the measured pumping speed against pressure on semilog paper.

QUESTIONS

The measurements you have just made may differ from those given by the manufacturer. What are some possible explanations?

Is the speed constant? Does it steadily increase? Does it steadily decrease? What reasons could you give for any variation in speed with pressure?

C. MEASUREMENT OF PUMPING SPEED OF PUMPING SYSTEM

Add accessory components (baffle, trap, valve, tubulations) between the diffusion pump and the vacuum chamber. As each component is added, measure the pumping speed at the vacuum chamber. If the speed is measured over the same range used with the pump by itself, interesting comparisons can be made.

QUESTIONS

What is the percentage loss in pumping speed as each component is added?

Is it about what you would expect from calculated measurements of component conductance?

Would you expect the percentage to be the same in each case if the order of assembly is changed?

D. PERFORMANCE OF THE PUMPING SYSTEM FOR VARIOUS GASES

Using the pumping system as assembled in the previous section, establish a constant pressure with the leak valve, using air as a leakage gas. Without changing the setting of the leak valve, cover the leak with different gases.

QUESTIONS

A change in gas will usually cause a change in indicated pressure. What are some contributory conditions?

How much change is due to the change in response of the gauge for this gas?

How much change is due to the difference in pumping speed for this gas?

1.2. OIL VAPOR DIFFUSION PUMP

BIBLIOGRAPHY

1. AVS Glossary of Terms Used in Vacuum Technology.

2. W.F. Brunner, Jr., and T.H. Batzer, <u>Practical Vacuum Techniques</u>, Reinhold, New York, 1965, pp. 36-39.

3. B.D. Power, <u>High Vacuum Pumping Equipment</u>, Chapman and Hall, London, 1966, pp. 45-55.

4. C.M. Van Atta, <u>Vacuum Science and Engineering</u>, McGraw-Hill, New York, 1965, pp. 227-240.

Experiment 1.3

CRYOSORPTION PUMP

W. F. Brunner and H. G. Patton

Lawrence Radiation Laboratory
Livermore, California

INTRODUCTION

The experiment describes operating principles and procedures for the cryosorption pump cooled by liquid nitrogen, the proper application of the pump over the pressure range in which it normally operates, and the limits imposed by both the gas species and the method of pump sequencing.

For this exercise two cryosorption pumps will be needed. Any commercial pump containing about three pounds of Zeolite will be suitable. However, with careful consideration adequate pumps can be constructed in most laboratories, as the construction is simple.

Although other cryogenic fluids, such as liquid helium and hydrogen are used to cool cryosorption pumps, the basic principles are not greatly different than with liquid nitrogen. This experiment is limited to liquid nitrogen.

OPERATING PRINCIPLES

The sorption pump consists of a vacuum-tight container filled with a sorbent, generally Zeolite or activated charcoal. The pump is chilled by immersion in a dewar of liquid nitrogen. Gas is expelled from the pump by increasing the temperature: a bakeout of 200°C is common practice.

Liquid nitrogen cryosorption pumps are normally used when elimination of an oil-sealed mechanical pump is desired. The cryosorption pump is quiet, clean, and vibration free and no external power is required for the pumping process. However, when properly prepared and applied the pump can be used as the primary pump when pressures as low as 10^{-5} Torr are desired. Physical adsorption is the process by which the pump operates. Quantities of gas are adsorbed onto surfaces by Van der Waals forces. The resident time for gas molecules is greatly increased by the cooling effect supplied by the liquid nitrogen. At the temperature of boiling liquid nitrogen (-196°C) low equilibrium pressures may exist at the sorbent surface. The pumping action of this pump is selective and only gases with liquifaction temperatures close to or higher than the sorbent temperature will be pumped significantly. This type of pump cannot exhaust to the atmosphere, which means that a limited amount of gas can be pumped before regeneration is required.

EXPERIMENTAL SYSTEM

Mount a Pyrex vacuum chamber on a test stand as shown in Fig. 1. Attach the cryosorption pump to the chamber with a section of vacuum line. A thermocouple gauge, to measure between atmosphere and 1 Torr, and a gas inlet valve should be attached to the top plate. Several bottled gases are required; helium, argon, nitrogen, and air. Other gases may be used: carbon monoxide, carbon dioxide, methane, etc. To prepare the pump for pumping, refer to the manufacturer's operating manual. If the pump is not a standard model, the recommendations given for standard pumps will usually apply. There are a variety of techniques used to activate pumps. In this experiment pump performance is being evaluated under several sets of circumstances; therefore, it is important to stay with one technique to avoid ambiguous results.

EXPERIMENTAL PROCEDURE

A. **SINGLE PUMP PERFORMANCE**

This set-up can be handled easiest if a small sorption pump is used. This is the size that contains approximately three pounds of

1.3. CRYOSORPTION PUMP

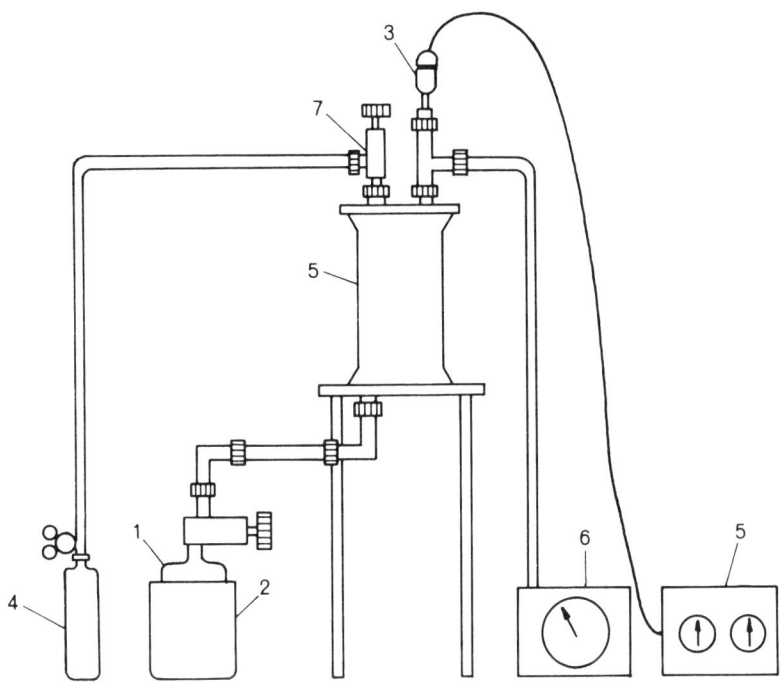

FIG. 1. Cryosorption pumped system. Equipment required: 1. cryosorption pump, 2. liquid nitrogen dewar, 3. thermal conductivity gauge, 4. sample gas bottle, 5. vacuum chamber, 6. vacuum gauge (760-1 Torr), 7. small vacuum valve.

molecular sieve and is described as being capable of handling 40-80 atmospheric liters of air.

For the first part of this exercise the performance of a single pump will be explored. The amount of gas pumped will be measured as well as the time to reach selected pressure levels.

Assemble the pump, gauges, gas bottle, and chamber as shown in Fig. 1. Bake out or prepare the pump in accordance with the manufacturer's recommended procedure.

The chamber should be filled to atmospheric pressure with one of the test gases by evacuation and back filling, or by flushing.

Record the initial pressure in the chamber. Open the valve on the cryosorption pump and record the time required to reach pre-selected pressures, 1 Torr, 10^{-1} Torr, 5×10^{-2} Torr, 2×10^{-2} Torr, 1×10^{-3} Torr, for example. When a base pressure has been reached, close the valve on the pump and fill the vacuum chamber with the test gas again. The apparatus is ready for the second evacuation cycle. Continue the cycles until the pump is saturated enough that significant pumping does not take place or until ten cycles have been completed. This test should be performed for each of the gases: argon, air, nitrogen, helium, etc.

Possible Hazards

When baking out the pump be sure the pump is equipped with a pressure relief valve or that the pump valve is open. Bursting pressures can be generated at bakeout temperatures.

QUESTIONS

What was the total amount of gas, in Torr-liters, pumped in ten runs for each gas?

What was the lowest pressure achievable for each gas?

Explain the differences in the amount pumped for each case.

What reasons can you give for any differences in base pressure?

What amount of gas could be pumped theoretically in each case?

B. MULTIPLE PUMP PERFORMANCE (PARALLEL)

For this part of the experiment the performance of two pumps connected in parallel are investigated.

Attach the two cryosorption pumps to the chamber as shown in Fig. 2.

Prepare these pumps in the same manner as before. Using both pumps in parallel, fill the chamber with the desired gas by back-

1.3. CRYOSORPTION PUMP

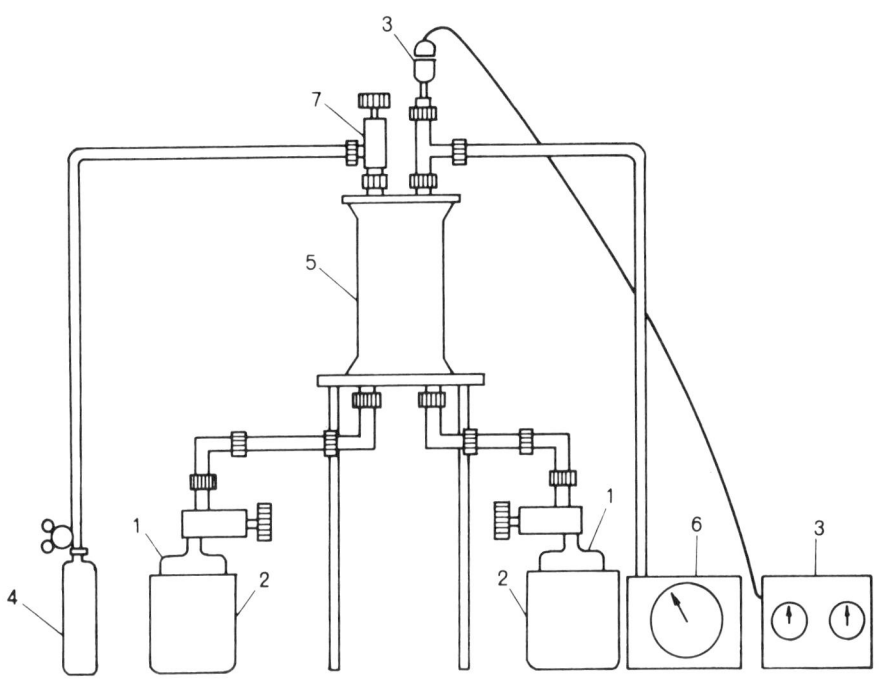

FIG. 2. Double cryosorption pumped system. Equipment required: 1. cryosorption pump, 2. liquid nitrogen dewar, 3. thermal conductivity gauge, 4. sample gas cylinder, 5. Pyrex vacuum chamber, 6. vacuum gauge (760-1 Torr), 7. small vacuum valve.

filling or flushing, as before. When ready to pump, open both pump valves simultaneously. Plot the length of time to reach selected pressure levels. Conduct the test in the same way for each gas.

QUESTIONS

Was the amount of gas pumped in ten runs doubled using this technique?

Was the pump down-time affected?

What explanation can you give for the variations between this test and the previous one?

C. MULTIPLE PUMP PERFORMANCE (SERIES)

Conduct this test using two cryosorption pumps attached to the vacuum chamber in series as shown in Fig. 2. Prepare the pumps in the same manner used in the two previous tests. Backfill or flush the chamber with the gas to be pumped. Open the valve to one pump only and pump the gas from the chamber until a pressure of 5 to 10 Torr is reached. Close the valve to this pump and open the valve to the second pump and allow it to reach base pressure. In actual practice the crossover pressure from one pump to another will be set by operating requirements and conditions. Observe the length of time to reach the preselected pressures. Repeat this test for ten cycles on each gas.

QUESTIONS

What was the total amount of gas pumped in ten runs?

What base pressure was achieved in each case?

What difference in pump down-time was observed between series and parallel operation?

DISCUSSION

Discuss the base pressures achieved, comparing parallel with series operation.

Explain which of the three pumping techniques would have the highest pumping speed.

Which technique would have the lowest base pressure?

Describe conditions in which a series pumping technique, parallel pumping technique, and a single pump technique would be used.

1.3. CRYOSORPTION PUMP

BIBLIOGRAPHY

1. B.D. Power, <u>High Vacuum Pumping Equipment</u>, Chapman and Hall, London, 1966, Chap. 8.

2. A.E. Barrington, <u>High Vacuum Engineering</u>, Prentice-Hall, Englewood Cliffs, N.J., 1964, pp. 125-142.

Experiment 1.4

GETTER-ION PUMP

W. F. Brunner and H. G. Patton

Lawrence Radiation Laboratory
Livermore, California

INTRODUCTION

This experiment describes operating procedures and principles for the getter-ion pump, the proper application of the pump over the pressure range in which it normally operates, and the limits imposed by the gas being pumped and the system condition.

The getter-ion pump functions without pump fluids; it has no moving parts. It is quiet and vibration free, and needs only electrical power for operation. It does not do well at pressures above 10^{-5} Torr, but will provide pressures less than 10^{-10} Torr without extreme difficulty. The pump must be evacuated initially to about 10^{-2} Torr before it will start properly. This is usually accomplished by trapped mechanical pumps or by cryosorption pumps. When the getter-ion pump is operating, the roughing pump is not needed and can be closed off from the chamber.

A predictable, stable pumping speed cannot be ascribed to getter-ion pumps in the same sense that is possible with diffusion or mechanical pumps. Pumping speeds quoted must be for a specific gas; and, even in this case, speeds may vary according to previous history, the condition of the pump, what other gases are present, and what type, etc.

In the getter-ion pump, gas is "pumped," that is, removed from the gas phase, by a combination of sorption and ionization. Pumping by sorption is accomplished by producing clean, active metal surfaces within the pump. Molecules of the active gases are removed from the gas phase by reacting with these surfaces. Pumping by ionization is accomplished by ionizing the gas molecules, then accelerating them to be buried in the pump surfaces. Molecules of both the active and inert gases are removed from the gas phase by this action. The active metal surface needed for sorption may be produced by either sputtering or evaporation. The ionization is accomplished by electron emission from either hot cathode or cold cathode surfaces. Many different types are manufactured; however, the most popular and the type available in the greatest range of size is the sputter-ion pump. This is the type used in this experiment. A 50-liter/sec pump will be most suitable, although a smaller pump may be adequate if the enclosure is appropriately smaller and the gas loads reduced. If the vacuum chamber has a minimum number of elastomer seals the results of the experiment will be made much clearer.

OPERATING PRINCIPLES

A schematic representation of a single-cell, diode-type sputter-ion pump is shown in Fig. 1. The anode is a tube with either a square or round cross section, suspended between and electrically insulated from the flat cathode plates, which are usually grounded to the pump housing. The magnetic field is parallel to the axis of the anode tube and normal to the cathode plates. The electrode material is usually titanium, the housing is made of stainless steel, the magnetic field is provided by a permanent magnet. Larger pumps are simply multiples of this single cell arranged to provide optimum opportunity for gas molecules to enter the cell cavity while taking maximum advantage of the available magnetic field.

In the sputter-ion pump the ionization by which both inert and active gas molecules are pumped is accomplished by cold cathode discharge. The impact and burial of the resultant ions on the cathode causes titanium metal to be ejected from the cathode (sputtered).

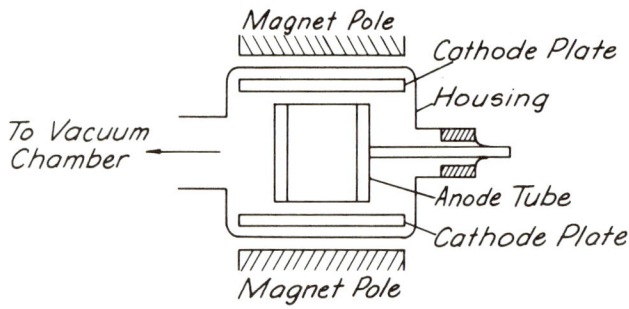

FIG. 1. Schematic representation of single-cell, diode sputter-ion pump.

1.4. GETTER-ION PUMP

The deposition of the ejected metal on adjacent pump walls provides the active metal surface needed for sorption.

This condition is created by establishing a potential difference of several thousand volts between cathode and anode. A random event, such as cosmic radiation, creates several ions in the residual gas. These are attracted to the cathode electrodes where, on impact, they cause the emission of electrons as well as titanium metal. The titanium condenses as a film on adjacent surfaces, the electrons move toward the anode but are forced by the magnetic field to take a helical trajectory, thus increasing the path length needed to sustain maximum ionization conditions.

EXPERIMENTAL SYSTEM AND PROCEDURE

This experiment has been developed for a 50-liter/sec sputter-ion pump, however, a pump of a different size or type will be quite adequate as the principles will be very much the same. A manufacturer's instruction manual for the pump selected can be of distinct advantage. Prior to attaching the pump to a vacuum chamber, inspect the interior to make sure it is free of oil films, loose flakes, or heavy deposits and that the cathode plates are not excessively etched. The pump can be attached to the chamber by bolting it directly to the base plate; this will make connecting tubing unnecessary. The vacuum chamber should be equipped with an ionization gauge, a thermal conductivity gauge, a sample gas inlet (leak valve), and a roughing valve. Since part of the experiment is to observe the ability of the pump to handle various gases, gases other than air are used--argon, nitrogen, oxygen, carbon dioxide, and helium, for example.

Assemble the pump and chamber with gauges as shown in Fig. 2. A cryosorption or well trapped mechanical pump can be used to rough out the ion pump and chamber. Evacuate the chamber to between 10^{-3} and 10^{-2} Torr with the roughing pump. Most modern getter-ion pumps start in this pressure range; if more specific pressure limits are required, they can be obtained from the manufacturer.

When the pressure has reached the starting range, turn the pump control switch on. Initially, the pump current will be quite high. It should begin to drop in a short period of time, however, depending on the size of the chamber and the pump, as well as their condition. If the high current condition persists, the fault can lie with the pump or vacuum chamber. If the fault lies with the pump, it could be due to excessive amounts of gas sorbed on the pump elements--usually caused by prolonged exposure to atmospheric conditions--or insulating films on the cathodes which inhibit sputtering, preventing the pump from pumping. The first condition can be rectified by prolonged

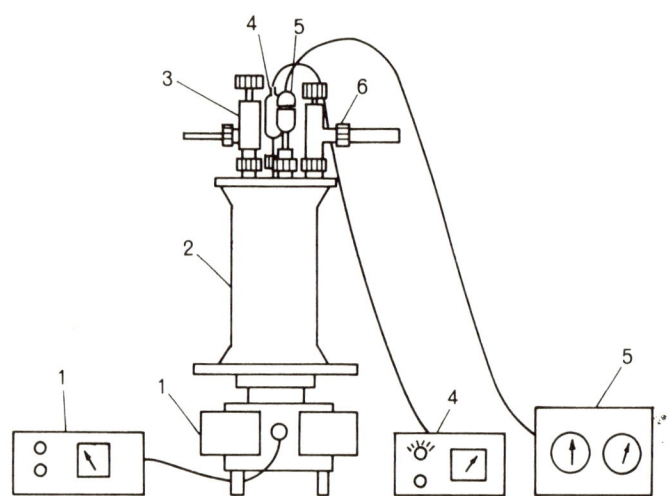

FIG. 2. Getter-ion pumped system. Equipment required:
1. getter-ion pump, 2. vacuum chamber, 3. leak valve, 4. ionization gauge, 5. thermal conductivity gauge, 6. roughing valve and pump.

pumping using the roughing pumps; the period of time can be reduced by raising the temperature of the pump, which accelerates the outgassing. The second condition can be corrected by chemical cleaning or, in less severe cases, flushing the pump with a gas producing a high sputtering rate, such as argon. If the fault lies with the vacuum chamber it will be due to an excessive gas load caused by inleakage or outgassing. Outgassing can be reduced by extended pumping with the roughing pumps or degassing by heating the chamber. In the case of inleakage, it is possible to locate the leak using the sputter-ion pump as a sensing instrument.

If the pump current begins to fall, the chamber pressure as indicated on the ion gauge should also fall. Allow the pump to reach its ultimate pressure; this may take several hours, or in more severe cases, overnight.

Open the leak valve and raise the pressure one or two decades with atmospheric air. (Do not exceed 5×10^{-5} Torr.) Allow several minutes for the condition to stabilize. Record the pump current and

1.4. GETTER-ION PUMP

the ion gauge reading. Cover the air leak with each of the test gases. Helium should be used sparingly to avoid saturating the pump with it. Record the pump current and ionization gauge reading for each gas.

Possible Hazards

The cable to the getter-ion pump carries high voltage. Care should be taken to see that the high voltage cable is in place, with the shield grounded, before the power supply is turned on.

QUESTIONS

Explain the variation in ionization gauge and pump current indications with the change in test gas. Consider the rate of flow through the leak, ionization probability, and pumping characteristics.

How does this information compare with the manufacturer's data?

BIBLIOGRAPHY

1. B.D. Power, High Vacuum Pumping Equipment, Chapman and Hall, London, 1966,

2. A.E. Barrington, High Vacuum Engineering, Prentice-Hall, Englewood Cliffs, N.J., 1964.

3. J.W. Ackley et al., Trans. 9th Nat. Vac. Symp, 380 (1962).

Experiment 1.5

VACUUM MEASUREMENT TECHNIQUES

W. F. Brunner and H. G. Patton

Lawrence Radiation Laboratory
Livermore, California

INTRODUCTION

Between atmospheric pressure and one Torr there is little difficulty in obtaining accurate measurements of the vacuum in an enclosure. The gauges used--Bourdon tube, diaphragm, liquid level-- are direct reading, that is, they measure the force exerted on a solid or liquid surface by the gases in the enclosure independent of the gas type.

From one Torr down to 10^{-3} Torr accurate measurement becomes more difficult, as the force becomes quite small and the standard form of gauges used above one Torr are not sufficiently sensitive. Special forms of these gauges that are adequately sensitive are available commercially, however, and accurate measurements can be made, but the gauges are delicate and measurements may be tedious and time consuming. In this range, there is another type of gauge that can be used, the thermal response gauge. This gauge measures the thermal conductivity of the gas in an enclosure. Although not precisely linear, the thermal conductivity of a gas varies directly with the pressure between one Torr and 10^{-3} Torr and can thus be used, when properly calibrated, to provide an indication of the amount of gas in the enclosure on which it is installed. It is a robust gauge and easy to use; the response to a change in the vacuum is an electrical readout, making it an ideal instrument for remote reading. It has a distinct disadvantage, however; the response is dependent upon the type of gas as well as the amount. Thus, the first order of difficulty in vacuum measurement is encountered. It is rarely simple to unambiguously define the gas type in the usual vacuum chamber, thus the usual procedure is to use a gauge calibrated for the thermal conductivity of nitrogen and express the measurement as a nitrogen-equivalent reading.

From 10^{-3} Torr down to 10^{-6} Torr, the situation becomes more complicated. Few of the direct-reading gauges are sufficiently sensitive, and those that are, such as the McLeod and capacitance diaphragm gauges, must be carefully set up. The method of use becomes a great deal more tedious and time consuming. In this range is another indirect-reading gauge, the ionization gauge. It measures the molecular density which, of course, is a direct measure of the amount of gas in the enclosure, but it does this by ionizing the gas and electrically "counting" the ions formed by collecting them on a charged electrode and measuring the small current produced. The probability that a gas molecule will be ionized varies with the type of gas. Thus the same general condition exists in this case that exists in the case of the thermal response gauge. It is difficult to define the gas type in the usual vacuum chamber. The generally accepted procedure is the same also, calibrate the gauge for nitrogen and express the measurement as a nitrogen-equivalent reading.

Below 10^{-6} Torr, the ionization gauge is the only type of gauge used except for special and extremely rare cases. In addition to the difficulty encountered in knowing the ionization probability of the gas being measured, the mere operation of the ionization gauge may produce a variation in the conditions within the enclosure. The problems involved in obtaining an accurate measurement of the amount of gas in an evacuated enclosure are many and complex; a subtle, but very important corollary to this is that an inaccurate measurement is rarely obvious, or even apparent. Fortunately, in the majority of cases it is only necessary to know the extent of the ambiguity to be expected from a particular type of gauge in a given situation. From this, an entirely adequate assessment of the conditions within an evacuated space can be obtained.

The operations suggested in this experiment cover the set-up and operation of each type of gauge mentioned. The procedures outline the methods by which comparisons of response from the various types of gauges can be made. Limitations on gauge functions and precautions to observe in their use are indicated. Methods for defining the accuracy of the gauges or for calibrating are outside the scope of this section, but several excellent sources for this information exist.

OPERATING PRINCIPLES OF MECHANICAL AND LIQUID LEVEL GAUGES

Mechanical and liquid level gauges measure pressure as a physical force exerted on a solid or liquid surface. Gauges of this

1.5. VACUUM MEASUREMENT TECHNIQUES

kind are available which can measure pressure from atmosphere to less than 10^{-5} Torr. In the liquid level gauge the pressure is measured as a difference in fluid level. The mechanical gauge measures pressure by the amount of deformation of a thin wall. The gauges chosen for this discussion are the most frequently encountered and demonstrate the operating principle; simpler and more complex forms may be encountered.

A. MECHANICAL GAUGES

A mechanical gauge consists of a hermetically sealed capsule with a thin wall. The thin wall deflects when a pressure differential exists on opposite sides of the wall. The deflection is transmitted and magnified by a mechanical linkage which causes a pointer to move across a dial face.

1. Bourdon Gauge

In the Bourdon gauge the capsule has the form of a tube slightly flattened into an elliptical cross section and shaped in the form of an arc. One end of the tube is fixed; the other end is free to move. The fixed end has a fitting that provides communications between the inside of the tube and the vacuum chamber; the free end is sealed and attached to the gauge mechanism. When the pressure in the vacuum chamber is reduced, the pressure in the tube is reduced and the tube radius decreases, causing the free end of the tube to move. The gauge mechanism is actuated and drives a pointer over a dial face.

When the inlet tubulation of the gauge is connected to the vacuum chamber, the inside of the Bourdon tube assumes the same pressure as the chamber. As the pressure in the chamber--and thus the pressure in the tube--is reduced, the tube radius decreases, causing the pointer on the dial to indicate the change in pressure. Atmospheric pressure is used as a reference; thus, in all but a few cases, the indication of the gauge is zero at atmospheric pressure, with the scale reading to 30 in. of mercury. This may cause some confusion in usage.

2. Diaphragm Gauge

The pressure sensing element of the diaphragm gauge is a sealed, cup-shaped capsule. One wall of the capsule is a thin, flexible diaphragm. Deflections in the diaphragm caused by pressure changes external to the capsule are transmitted through a mechanical linkage to the pointer, which moves across a dial face.

The inlet fitting of a diaphragm gauge such as the Wallace and Tiernan is connected to the hermetically sealed case of the instrument. A sealed, evacuated capsule inside the case provides the reference pressure. As the pressure in the case is reduced, the thin diaphragm of the capsule extends and the gauge pointer moves from 760 Torr toward zero.

B. LIQUID LEVEL GAUGES

A liquid level gauge measures pressure as a difference in liquid level. The gauge liquid is held in a specially shaped container (a U-shaped tube is an example of a common type). By the shape of the container, the body of the liquid is kept continuous while portions of its surface are separated. If the surfaces are at a common level, the pressure on each surface is identical. If the surfaces are at different levels, the vertical distance between them represents the difference in pressure between them. Obviously, the upper level is the lower pressure. If the difference in vertical height is measured in millimeters and the gauge liquid is mercury, the pressure difference can be read in Torr, as each millimeter is, for all practical purposes, equal to one Torr. With any other liquid, the measurement must be multiplied by the ratio of the density of the liquid to the density of mercury to get the pressure difference in Torr.

1. U-tube Gauge

The U-tube gauge is the clearest and simplest example of a liquid level gauge. It consists of a glass U-tube partially filled with a liquid. The usual liquid is mercury because it does not readily "wet" the glass and less time is required to wait until the gauge liquid drains from the wall. The vapor pressure of mercury is low at room temperature (10^{-3} Torr) compared to the pressure range being measured, and the readout can be in Torr without conversion. To use this gauge, one arm of the U-tube is connected to the region where the pressure is to be measured, while the pressure above the surface in the other arm is held at some stable reference value. The lowest pressure that can be measured with the standard form of this gauge is determined by the ability to read the difference in height, or about 0.5 Torr.

The usual form of U-tube gauge used for vacuum measurements is the closed-end type. In this case one arm of the U is closed and evacuated; the open end is attached to the vacuum chamber. If the U-tube is long enough (about 80 cm) there will be an evacuated space above the mercury in the closed end. This is the so-called Torricellean vacuum. The mercury will be 760 mm lower in the open end at standard atmospheric pressure. When the pressure in the vacuum chamber is reduced, the new pressure in Torr is simply the difference in levels, measured in millimeters, of the two mercury columns.

1.5. VACUUM MEASUREMENT TECHNIQUES

2. Dubrovin Gauge

The Dubrovin is a float type gauge; its construction is shown in Fig. 1. A steel cylinder of small diameter, closed at one end, floats vertically in mercury with its open end submerged. The air is evacuated from the chamber formed by the floating cylinder. The outer glass tube is connected to the vacuum system through a stopcock. A scale with an adjustable mounting is attached to the frame of the gauge just outside the outer glass tube. By sighting along a pointer attached to the top of the floating stainless steel tube to this scale, the pressure in the gauge can be determined.

The operating principle of the gauge is as follows: When both the float chamber and the gauge chamber have been evacuated, the bouyant force acting on the float (the weight of mercury displaced by its submerged walls) is in equilibrium with the weight of the float. Under these conditions the float is in its highest position and the pointer should indicate zero. When the pressure in the outer chamber increases, it exerts a downward force in the float, causing it to sink until the additional mercury displaced is sufficient to compensate for the downward force and again produce equilibrium. The pointer indicates the pressure being exerted on the float. Note that the linear distance from the positions marked 0 and 20 mm on the scale is much more than 20 mm. The movement of the float has "magnified" the liquid-level effect. The "magnification" achieved is a function of the physical dimensions of the float chamber.

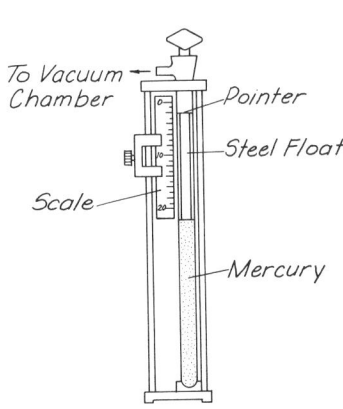

FIG. 1. Dubrovin gauge.

Correct operation of the gauge depends on the float chamber being evacuated to a low pressure, 0.1 Torr or less. This can be accomplished by laying the gauge on its side so the float chamber is exposed to the gauge chamber and evacuating both chambers to 0.1 Torr or less. While still evacuated, the gauge is returned to a vertical position with the float chamber floating in the mercury. If, at this time, the pointer on the float does not indicate zero, the scale should be adjusted until it does. Now, if the pressure in the gauge chamber is increased, the float will be pushed down deeper in the mercury, the pointer on the top of the float indicating the change in pressure on the scale.

3. McLeod Gauge

In this gauge, the liquid-level principle is extended to lower pressures. This is accomplished by capturing and isolating a sample of the low pressure gas in a known volume, then compressing it into a much smaller known volume. According to Boyle's Law, this will result in a proportional increase in pressure which, with appropriate gauge design, can be measured as a difference in liquid level. Now, since the volume of the sample at this pressure is known, the initial pressure of the sample can be calculated using Boyle's Law, where

$$P_1 V_1 = P_2 V_2 \text{ and } P_1 = P_2 \frac{V_2}{V_1}$$

Thus, if the initial volume of the isolated sample is V_1; the initial pressure of the isolated sample is P_1; the smaller, known volume is V_2; and the measured final pressure is P_2, then the initial pressure can be obtained by multiplying the difference in liquid level by the ratio of the two known volumes.

The essential elements of a typical McLeod gauge are shown in Fig. 2. The inlet tube connects the vacuum chamber with the bulb. The bulb is the volume in which compression, by which the gauge functions, takes place. The gas in the bulb is compressed into the closed capillary where the final volume is defined. The final pressure of the compressed gas in the closed capillary is determined by the difference in liquid level between it and the closed capillary.

There are many methods for raising the mercury in the gauge. One of the most popular is to allow atmospheric pressure to force the mercury up into the gauge and use an auxiliary vacuum to lower it. This is the method used in describing McLeod gauge operation for this experiment.

1.5. VACUUM MEASUREMENT TECHNIQUES

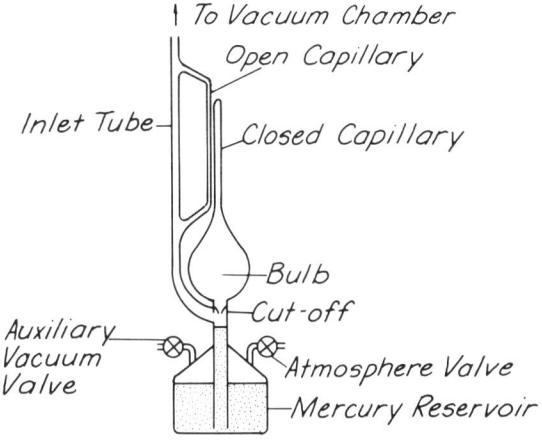

FIG. 2. McLeod gauge.

To obtain a pressure measurement, the mercury level is raised by opening the atmosphere valve. As the mercury level rises, it passes the bulb cut-off and isolates the bulb from the vacuum chamber. It then begins to act like a liquid piston in the bulb, compressing the gas into the small volume of the closed capillary. The mercury level can be stopped by closing the atmosphere valve when both levels are within the parallel section of the capillaries, where the difference in level can be measured. The pressure above the mercury level in the closed capillary is the final pressure; its value is obtained from the difference in level between the closed and comparison capillaries. The volume above the mercury level in the closed capillary is the final volume; its value is obtained by multiplying the distance between the mercury level and the top of the capillary by the cross-sectional area of the capillary. The initial volume is the combined volume of the bulb from the cut-off to the top of the closed capillary. When the measurement is completed, the mercury level is lowered below the cut-off by opening the auxiliary vacuum valve. Before another measurement is taken some time must be allowed for the pressure in the bulb to come to equilibrium with the pressure in the vacuum chamber.

EXPERIMENTAL PROCEDURES

A. COMPARISON OF THE BOURDON AND U-TUBE GAUGES

Attach a gauge manifold to a mechanical pump, placing a valve between the pump and the manifold. On this manifold, mount a Bourdon gauge and a U-tube gauge.

Note the reading of the gauges at atmospheric pressure, then open the valve and reduce the pressure as indicated by the Bourdon gauge to 10 in. Record the reading of the U-tube at this level. Reduce the pressure to 15 in. and note the U-tube reading. Reduce the pressure to 20 in. and again note the U-tube reading.

Continue pumping until no further decrease in gauge reading is discernable. Record the pressure indicated by each gauge.

Vent the manifold to atmospheric pressure.

Possible Hazards

The mercury U-tube must be vented slowly to avoid the liquid hammer on the closed end, which could break the gauge and cause a mercury spill.

QUESTIONS

What are some of the reasons for the difference in readings of the Bourdon and U-tube gauges?

Would you expect the two gauges to read the same when further pumping produces no further decrease in gauge reading?

B. COMPARISON OF THE U-TUBE AND DUBROVIN GAUGES

Remove the Bourdon gauge from the manifold, leave the U-tube, and install the Dubrovin gauge. Open the valve and evacuate the manifold.

Compare the readings of the two gauges over the range of the Dubrovin.

1.5. VACUUM MEASUREMENT TECHNIQUES

Vent the manifold to atmospheric pressure.

QUESTIONS

Do the readings of the two gauges agree?

If not, is the error constant? What kind of error would this indicate?

If not, is the error uniformly changing? What kind of error would this indicate?

C. COMPARISON OF THE DUBROVIN AND McLEOD GAUGES

Remove the U-tube from the manifold, leave the Dubrovin, and install the McLeod gauge. This installation should include a trap between the McLeod gauge and the manifold to keep mercury vapor out of the remainder of the vacuum system and to reduce the amount of condensable vapor entering the McLeod gauge. Before evacuating this manifold, some thought should be given to procedure, as the evacuation of a McLeod gauge must be done carefully. There is very little friction between mercury and glass; thus the mercury will surge rapidly through the gauge if changes are made too quickly. This situation is very undesirable due to the fragile nature of the glass construction and the heavy mass of mercury being moved about.

When the McLeod gauge is evacuated, the mercury level will rise as the pressure in the gauge is reduced. With a McLeod gauge that uses an auxiliary vacuum to manipulate the mercury, the auxiliary vacuum valve is opened to reduce the pressure on the surface of the mercury in the reservoir at the same rate as the pressure in the gauge is being reduced. When the gauge is evacuated, both the atmosphere valve and the auxiliary vacuum valve are closed and the gauge is ready to be used. If it is possible to use a valve on the inlet of the gauge, the gauge can be kept under vacuum when not in use. This would make it easier to put into use; it would also keep both gauge elements and mercury clean.

If the gauge and manifold are to be pumped down together, the rate of evacuation can be controlled by the mechanical pump valve. The rate of pressure decrease on the mercury reservoir is controlled by the auxiliary vacuum valve. Done properly, the mercury will remain at a constant level without violent bubbling. When the pressure in the manifold is about 10 or 15 mm of mercury as indicated by the Dubrovin gauge, close the pump valve and the auxiliary vacuum valve. Allow several minutes for conditions in the gauges and the manifold to stabilize. Then open the atmosphere valve on the mercury reservoir slowly so the level will begin to rise. Just before the level reaches the cut-off, note the reading of the Dubrovin gauge. The rise of mercury level can be stopped when it reaches the desired point by closing the atmosphere valve. The level should be stopped when it reaches the point indicated in the gauge operating instructions for the pressure range to be measured.

Make at least three more measurements, reducing the pressure five Torr or so each time.

QUESTIONS

Compare the McLeod gauge reading with the reading of the Dubrovin. Does it read the same? Would you expect it to?

If there is a difference, what could be contributing to it?

OPERATING PRINCIPLES OF THERMAL CONDUCTIVITY GAUGES

At pressures below one Torr, the transfer of thermal energy through a gas by conduction decreases linearly with the pressure. If an electrically heated wire is suspended in an evacuated tube it will lose heat to the wall of the tube at a rate dependent on the gas density (pressure) there. If the wire is heated by a constant power source, the wire will be hot when the pressure is low and become cooler as the pressure increases. With proper calibration, the temperature of the wire can be used to measure the relative pressure in the tube. This is the principle and general function of the thermal conductivity gauge.

Thermal conductivity gauges will measure the total pressure of both permanent gases and vapors, but the calibration will vary with each type. The response for a given gas type will vary too, if the condition of the wire surface changes due to an accumulation of foreign matter.

1. The Thermocouple Gauge

In the thermocouple gauge, the filament wire temperature is measured with a thermocouple junction. The junction is spot-welded

1.5. VACUUM MEASUREMENT TECHNIQUES

to the heater wire to ensure good thermal transfer. The voltage developed by the junction (about 5-15 mV in most cases) is registered on a millivoltmeter calibrated in Torr or microns of mercury. The heater current varies with the manufacturer but usually lies between 10 and 150 mA. The voltage to the heater wire also varies; it may be ac or dc, and anywhere from 5 to 115 V. Battery powered units are not unusual. The thermocouple gauge is simple, inexpensive, and rugged. However, it has a nonlinear scale and thus its use is generally limited to situations where accuracy is not required.

Thermocouple gauge tubes may be made of either metal or glass; they are usually about 1-1/4 in. in diameter and between 2 and 6 in. long. The metal tube is by far the most popular, since the sturdy nature of the gauge elements makes it a preferred choice for installations where the breakage risk is high. The inlet of the tube is usually a short section of pipe with a 1/8 NPT thread by which the tube is attached to the vacuum enclosure. A sealant is used on the threads to make the joint leak-tight. Thermocouple gauge tubes of different manufacturers are rarely interchangeable. Even when a tube is being replaced by another tube of the same model and by the same manufacturer, the replacement tube should be calibrated or a large error may result.

Putting the gauge into operation is quite simple. The tube is attached to the vacuum enclosure and the cable is installed between tube and power supply. The gauge can be turned on at atmospheric pressure, since the wire temperature is not high enough (usually about 150-250°C) to be harmful. If the heater current is known, this value (obtained from the operating instructions) can be set at atmospheric pressure; however, it is advisable to recheck it at a lower pressure to be sure it has not changed. If the heater current is not known, the proper setting can be obtained by reducing the pressure in the gauge tube to a value well below its lower limit and adjusting the current until the output meter indicates zero.

2. The Pirani Gauge

In the Pirani gauge, the change in filament wire temperature is determined by a change in its electrical resistance. The resistance is measured with a Wheatstone bridge network. Most Pirani gauge tubes contain two identical wire filaments, one is in a capsule sealed off at a pressure less than the lower limit of the gauge (the dummy element). The other is in the same kind of capsule, but the capsule is open to the gauge tube (the gauge element). By connecting the two in separate arms of the bridge (see Fig. 3), fluctuations in bridge balance due to changes in ambient temperature and supply voltage tend to be compensated. The Pirani gauge circuit is a little more complex than the thermocouple and therefore not quite as trouble free. The zero setting has a tendency to drift, making it necessary to check the output periodically. The gauge elements are not as rugged as those of the thermocouple but the scale is nearly linear, which makes it the preferred choice for reliable measurements.

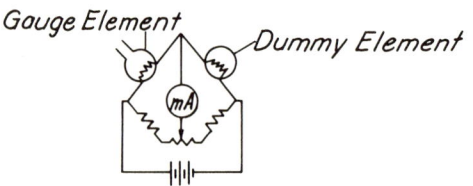

FIG. 3. Pirani gauge.

Pirani gauge tubes may be made of either metal or glass; the size varies widely with the manufacturer. The inlet is usually a tube about 3/8 in. in diameter, which is sealed to the vacuum enclosure with a gasketed compression fitting. Pirani gauge tubes of different manufacturers are not interchangeable. When a gauge tube must be replaced, the replacement must be calibrated to the power supply before it is used or it will not provide reliable measurements.

Putting the gauge into operation is simply a matter of sealing the tube to the vacuum enclosure, connecting the cable from tube to power supply, and turning the power supply on. There is no simple way to calibrate the gauge tube to the power supply while it is installed. The calibration can be <u>tested</u>, however, by reducing the pressure in the tube to 10^{-3} Torr or less. If the indication is less than zero or more than zero, the calibration should be checked.

EXPERIMENTAL PROCEDURES

THERMAL CONDUCTIVITY GAUGE OPERATION AND COMPARISON

Set up a high vacuum pumping station capable of providing a speed of at least 10 liter/sec at 10^{-4} Torr or less. The small pumping station described in Experiment 1.2, consisting of an oil vapor diffusion pump, cooled baffle, and valve is quite adequate for this purpose. Install a small vacuum chamber on this pumping station. The chamber must be equipped with feedthrough fittings to which the

1.5. VACUUM MEASUREMENT TECHNIQUES

gauges to be operated and compared can be attached. This chamber can be a short length of 6-in. i.d. Pyrex pipe and a cover plate fitted with appropriate feedthroughs.

Install a thermocouple gauge tube, a Pirani gauge tube, a Dubrovin gauge, and a McLeod gauge on the vacuum chamber. A vapor trap should be installed between the McLeod gauge and the chamber. If, in addition, a valve is installed between the trap and the chamber, the hazards of violently agitated mercury are lessened and the frequent venting and pumpdown required for this experiment are made less tedious and time consuming.

Read the operating instructions for each gauge over carefully and perform any pre-pumpdown checks that are indicated, then evacuate the chamber to 1 Torr as indicated on the Dubrovin gauge. Check this with the McLeod gauge and compare it with the indications of the thermocouple gauge and the Pirani gauge.

QUESTIONS

Under what conditions could a variation in the pressure measured by the Dubrovin and McLeod gauge be valid?

Which of these two gauges would you expect the thermal conductivity gauge to agree with?

Do the thermocouple and Pirani gauges agree? Is there a condition by which a valid difference could exist in this case?

Backfill the chamber to 1 atm with nitrogen, re-evacuate to 1 Torr and recheck. If the variations disappear what could you tell about the previous conditions?

Evacuate the chamber to a pressure less than 10^{-4} Torr as measured with the McLeod gauge. If the thermocouple and Pirani gauges do not read correctly, warm the tubes with a heat lamp or other mild heat source to about 50°C to accelerate tube outgassing, allow them to cool, and recheck. If they now read correctly, the next phase of the experiment can be started. If they still do not read correctly, additional pumping and outgassing should be tried. If this is unsuccessful, consult the gauge operating instructions for recalibrating procedures and recalibrate.

When the thermocouple and Pirani gauges are reading correctly at both the lower and upper limits, an evaluation of the variation in pressure indication with a change in gas can be made. With a leak valve, establish a pressure of 1 x 10^{-2} Torr in the chamber. It may be necessary to partially close the diffusion pump inlet valve to get this pressure without raising the backing pressure at the diffusion pump outlet excessively.

Tabulate the indications of each of the gauges. Cover the leak with nitrogen and tabulate this reading also. Continue covering the leak with various gases and tabulating the response. The effect of acetone and water vapor should be interesting to check also.

Repeat this procedure at several pressure levels and plot a family of curves showing thermal conductivity gauge response compared to McLeod gauge indications.

QUESTIONS

Do the thermocouple gauge and the Pirani gauge agree?

If they are different, are they different by the same amount all the way, or does the difference vary with pressure?

OPERATING PRINCIPLES OF IONIZATION GAUGES

At pressures of 10^{-3} Torr and below there are not enough gas molecules present to supply a substantial indication of pressure by mechanical means or by thermal response. For this pressure range, the measuring instrument most commonly used is the ionization gauge. The measurement takes place in a gauge tube attached to the vacuum system and is accomplished by converting the gas molecules in the vicinity of the gauge tube elements into positively charged ions. These ions are collected on a negatively charged electrode, which "counts" them by measuring the amount of current they produce when they hit the electrode.

The conversion of gas molecules into ions is accomplished by beaming high velocity electrons through the gas. An electron hitting a gas molecule knocks an electron off the molecule, leaving it with a positive charge. For a constant current of electrons, accelerated over a given potential, the rate at which positive ions are formed will be directly proportional to the number of molecules per unit volume (gas density) in the path of the electron stream. Therefore, the ionization gauge with proper calibration can be used to measure pressure, since the pressure of a gas, assuming a constant temperature, is proportional to its density.

1.5. VACUUM MEASUREMENT TECHNIQUES

The ionization gauge can be calibrated by comparing its response to the reading obtained from an absolute instrument such as the McLeod gauge under a controlled set of conditions. However, even under the same conditions of gas density and electron energy, there will be a difference in the amount of ion current obtained from different gases (the ionization probability of the gas it is being calibrated with must be known). The ability to define the pressure by measuring the ion current is therefore limited to the ability to identify the gas being measured and to define its cross section for ionization. In most applications, the identification of gas type is impractical. The usual practice is to calibrate the gauge with nitrogen and indicate, when citing measurements, that they are "nitrogen-equivalent" readings.

There are several types of ionization gauges and various hybrids of these; the major difference between them is in the way the gas molecules are converted to ions. This experiment is confined to the two types that are encountered most frequently: the thermionic, or hot cathode gauge and the Philips, or cold cathode gauge.

1. The Thermionic Ionization Gauge

The conventional thermionic ionization gauge tube is made of glass and contains three electrodes; a cathode, a grid, and a collector. The cathode is a wire or ribbon filament heated to emission temperature by about 10 V at 3 to 4 A. The grid is a wire helix, approximately 20 mm in diameter, with about 5 mm spacing between turns; it is provided with a fixed potential of around 150 V positive with respect to the cathode. The collector is a metal tube about 40 mm in diameter surrounding the grid (it may also be a metal coating on the inside of the glass tube) and has a fixed potential of some 20 to 30 V negative imposed on it.

In operation, electrons emitted from the hot filament are attracted to the grid. Most of them miss the grid wire the first time, continue toward the collector and are turned back toward the grid by its negative potential. They oscillate back and forth in this way several times, creating a favorable condition for ionization of gas molecules. The positively ionized gas molecules are captured by the collector, each providing a small pulse of current which is registered on a microammeter placed in the circuit. The microammeter is marked off in pressure units. The upper limit of this gauge tube is about 1×10^{-2} Torr, its lower limit about 10^{-7} Torr.

The Bayard-Alpert type of thermionic ionization gauge has the same three electrodes; however, the positions of the filament and collector are reversed. The filament, heated by the same general power requirements, is mounted outside the grid. The grid has roughly the same relative potential and geometry as the conventional type. The collector is a wire running along the central axis of the grid. The function of this type is very much similar to the conventional one;

however, the upper limit in this case is about 10^{-4} Torr, its lower limit about 10^{-10} Torr.

The circuitry of the power supply for the thermionic ionization gauge is moderately complex. It must provide a variable voltage source for the filament and a means of regulating and reading emission current. It must also provide fixed voltages for both grid and collector and a means of reading the very small positive ion current.

Most suppliers provide operating instructions for their gauges which include recommended starting and operating procedures as well as schematics of the power supply circuit. An emission setting is given for the gauge tube supplied with the unit. It will be adequate for most cases, but will rarely be precisely correct. Most thermionic ionization gauge control units will accept more than one type or model of gauge tube. When interchanging, the values for a tube should be checked so the control can be adjusted to accomodate it.

To put a thermionic ionization gauge in operation, the tube is connected to the vacuum chamber with some kind of compression fitting. With the control power off, the control cable is attached. Some units have oriented plugs that will fit only one kind of tube, some simply have slip connectors on each conductor so any tube can be connected. When the pressure in the chamber is below the upper limit of tube operation, the control can be turned on. (It may be necessary to have a thermal conductivity gauge on the chamber to define when the pressure is within the limit.) The sequence for putting the tube in operation is usually slightly different for each control and is contained in the operating instructions for the control; but, if the following general conditions are established, most gauges will function reasonably well.

1. With power off, the mechanical zero of the ion current meter (pressure readout) is checked and reset if off.

2. With power on and filament emission off, the meter is again zeroed, this time electronically, which is in effect zeroing the electrometer.

3. With power on and filament on, the emission current is set.

4. The gauge is now functioning; the filament should be incandescent, and the range switch can be moved to the position that will allow the meter indication to be on-scale.

Any degassing of the gauge tube should be approached carefully and cautiously, with close reference to the operating instructions. The practice of degassing at pressures above 10^{-5} Torr is questionable; a gentle warming of the gauge tube envelope and a mild heating of the electrodes is all that should be attempted.

1.5. VACUUM MEASUREMENT TECHNIQUES

2. The Philips Cold Cathode Ionization Gauge

The configuration of the cold cathode gauge tube is quite simple. It consists of two cathode plates parallel to and opposite each other, with a ring-shaped anode suspended between them. The plane of the ring is parallel to the plane of the cathode plates. A potential of from 2000 to 4000 V dc is applied to these electrodes and a magnetic field of about 500 G is provided normal to the cathode surfaces. Electrons are emitted from the cathode surfaces due to the high voltage. The electrons are attracted to the anode, but in their flight toward it are caused to move in spiral paths due to the magnetic field. They travel back and forth in helical paths between the cathodes many times before striking the anode. This extended path length increases the probability that ionizing collisions will occur. The positive ions are of much higher mass than the electrons and therefore move directly to the cathodes. The total discharge current, which is the sum of the positive ion current to the cathodes and the electron current from the cathodes, is used to measure the pressure. The Philips cold cathode gauge does not provide precise measurements even if carefully calibrated, because the discharge characteristics are slightly unstable. However, it is entirely reliable as a pressure indicator for vacuum process applications, because of the sturdy nature of the tube elements, and is most frequently used in that type of service.

The typical commercial cold cathode gauge has a metal tube with the walls of the tube shaped to act as cathode surfaces. The anode, usually a wire loop or a short cylinder, is supported between them by the high voltage lead-in installed in the end of the tube. The insulator of the lead-in is shielded from conductive deposits of sputtered cathode metal which could give rise to erroneous pressure readings. The required magnetic field is supplied by a permanent magnet. The power supply for the gauge is extremely simple. It consists of a source of high voltage dc, a current-limiting resistor for safety, and a microammeter marked off in pressure units to indicate gauge response. A shielded, single conductor, high voltage cable connects the tube to the power supply. The shield should be grounded to both supply and tube.

Putting the gauge into service is elementary. The gauge tube is attached to the vacuum chamber, and with the power off, the high voltage cable is connected between the power supply and tube. It is important, for both safety reasons and for gauge performance, that a good ground exists between tube and power supply; the cable shielding is usually used for this purpose. When the vacuum chamber is pumped down, the power supply can be turned on and the range selector used to bring the meter indication on-scale so that a reading can be obtained. The gauge is not damaged by short term operation at atmospheric pressure, but extended operation at pressures higher than 5×10^{-2} Torr is undesirable.

EXPERIMENTAL PROCEDURES

IONIZATION GAUGE OPERATION AND COMPARISON

Set up a high vacuum pumping station. The one used for the thermal conductivity gauges will also be quite adequate for this portion of the exercise. Install a thermionic ionization gauge tube (use a conventional tube and a Bayard-Alpert tube if both are available), a Philips gauge tube, and a McLeod gauge on the chamber. Use a vapor trap and an isolation valve between the McLeod gauge and the chamber as before.

Read the operating instructions for each gauge over carefully and conduct any pre-pumpdown checks that are indicated.

Evacuate the chamber to the ultimate pressure of the system and check the operation of the gauges. Degas the tubes (mildly) and check their reading with the McLeod gauge.

QUESTIONS

How can you tell whether the gauges are clean and are reading properly?

What is the relationship of electron current to ion current?

Make several successive pressure readings with the McLeod gauge. Is the indication of the McLeod gauge changing? If so, what about the indication of all the other gauges?

If the McLeod gauge is not changing, are the others?

What conditions might be the basis for either of these situations?

Raise the pressure to 10^{-4} Torr as indicated on the conventional thermionic ionization gauge by admitting nitrogen to the vacuum chamber with the leak valve. Allow several minutes for the pressure to stabilize, then compare the readings of all the gauges. Record these readings and adjust the leak valve until the pressure drops to 8×10^{-5} Torr. Wait several minutes and record these readings. Continue this procedure, taking readings at 6×10^{-5} Torr, 4×10^{-5} Torr, 2×10^{-5}

1.5. VACUUM MEASUREMENT TECHNIQUES

Torr, and 1×10^{-5} Torr. Tabulate these readings, then plot them using a linear scale. Plot the McLeod gauge reading on the ordinate and the ionization gauge readings on the abcissa.

QUESTIONS

Do the plotted points for each gauge fall in a straight line?

If the gauges are reading properly, should they fall in a straight line? Explain.

What is the significance of the straight line plot and of a plot with a curved line?

Try the same procedure that has just been used with nitrogen, using other gases. To obtain a more complete exchange of gas, first pump the chamber to a pressure near the ultimate of the system, then admit the gas to be used, adjusting the leak valve until the pressure reaches 10^{-4} Torr. Allow the pressure to remain at 10^{-4} Torr several minutes, then close the valve and let the pressure return to near the system ultimate. Do this at least twice, then repeat the gauge comparison procedure as it was used with nitrogen.

QUESTIONS

Do all the gauges react in the same way to the replacement gas?

Should they react in the same way?

Should the gauge tubes be degassed before each comparison to read more correctly?

BIBLIOGRAPHY

1. J.H. Leck, *Pressure Measurement in Vacuum Systems*, Chapman and Hall, London, 1964.

2. A.E. Barrington, <u>High Vacuum Engineering</u>, Prentice-Hall, Englewood Cliffs, N.J., 1964.

3. C.M. Van Atta, <u>Vacuum Science and Engineering</u>, McGraw Hill, New York, 1965.

4. P.A. Redhead, J.P. Hobson, and E.V. Kornelson, <u>The Physical Basis of Ultrahigh Vacuum</u>, Chapman and Hall, London, 1968.

5. S. Dushman, <u>Scientific Foundations of Vacuum Technique</u> (J.M. Lafferty, ed.), 2nd edition, Wiley, New York, 1962.

Section 2

EXPERIMENTS WHICH ILLUSTRATE
THE CHARACTERISTICS OF THE VACUUM ENVIRONMENT

Experiment 2.1

DEMONSTRATION OF THE OUTGASSING
OF DIFFERENT VACUUM MATERIALS

F. Rosebury

Research Laboratory of Electronics
Massachusetts Institute of Technology
Cambridge, Massachusetts

INTRODUCTION

Materials used in the construction of vacuum systems and vacuum device components can act as sources of gas under a variety of conditions. When heated, all materials give off gas which may be due to any of several factors or a combination of these:

1. Atmospheric gases and vapors, especially water vapor, adsorbed, absorbed, or dissolved on or in the material

2. Chemical decomposition of the material, some of whose breakdown products may be gases and/or vapors

3. Evaporation of the material itself without decomposition

4. Permeation of gases through a material when it forms a vacuum envelope or enclosure

5. Foreign substances on the material; such contamination may be subject to 1, 2, and 3 above

It is to be noted that such evolution of gas generally increases as the material is heated in vacuum.

The simple experiment described here is designed only to demonstrate the outgassing properties of materials, and as a guide to the selection of such materials to be used in the construction of vacuum equipment. It is not intended as a method of measuring outgassing rates. Organic materials (e.g., plastics, greases, elastomers, etc.) give off vapors of different chemical composition at ambient or elevated temperatures. Some materials, such as plastics, may polymerize, and some metals, such as brass (which loses zinc because it has a higher vapor pressure than copper) change in composition under the action of heat.

EXPERIMENTAL SYSTEM

A conventional vacuum system is required consisting of a forepump, an oil- or mercury-diffusion pump (or ionization pump), a cold trap with liquid nitrogen refrigeration, and an ionization or other suitable vacuum gauge.

A small specimen of the material under investigation is placed in a hard glass tube with a constriction for sealing off, as illustrated in Fig. 1. A small tubular electric oven, controlled with a variable transformer such as a Variac, is used to heat the sample. Similar ovens are used to bake out the vacuum gauge and the cold trap between runs, so as to remove any condensed material. Temperatures can be measured with chromel-alumel thermocouples.

All samples are to be previously cleaned in a manner appropriate to the material (do not use acetone or other volatile solvents on

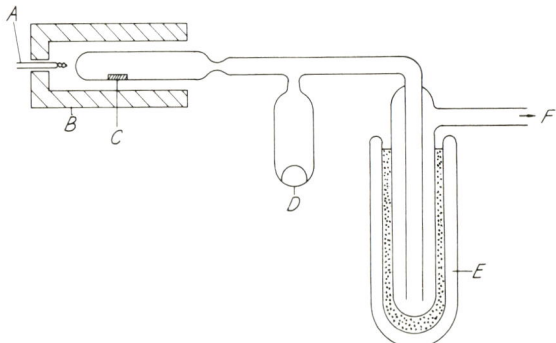

FIG. 1. Apparatus to demonstrate the outgassing of vacuum meterials. A. thermocouple, B. oven, C. specimen, D. vacuum gauge, E. cold trap, F. to pumps.

2.1. OUTGASSING OF VACUUM MATERIALS

plastics or elastomers; these are best cleaned in mild detergent-water solution in an ultrasonic bath).

EXPERIMENTAL PROCEDURE

The following observations are to be made:

1. The best initial pressure before and after prolonged pumping (several hours) with the specimen at room temperatures.

2. Heat is applied to the specimen, up to a maximum depending on the material (for organic materials this should obviously be below the decomposition or polymerization temperature) for a time sufficient for the pressure to reach a stable value. Notation is to be made of the elapsed time, the temperature, and the pressure at various times.

3. Heat is shut off and the terminal pressure is observed after prolonged pumping at room temperature or until there is no appreciable further reduction in pressure.

COMMENTS

1. It would be instructional to experiment with various materials and with the use of different analyzing procedures.

2. It can be clearly shown, by using different cleaning treatment on a material, that these techniques can dominate the outgassing characteristics. Differences in metals can be completely masked by surface cleaning procedures.

BIBLIOGRAPHY

1. N.K. Adam, *The Physics and Chemistry of Surfaces*, 3rd edition, Oxford University Press, London, 1941.

2. V.O. Altemose, *J. Appl. Phys.*, **32**, 1309 (1961).

3. R.M. Barrer, *Diffusion in and Through Solids*, Cambridge University Press, New York, 1941.

4. S. Dushman, *Scientific Foundations of Vacuum Technique* (J.M. Lafferty, ed.), 2nd edition, Wiley, New York, 1962.

5. R.E. Honig, *RCA Review*, 23, 567 (1962).

6. L.D. Jaffe and J.B. Rittenhouse, *A.R.S. Journal*, **32**, 314 (1962).

7. W.H. Kohl, *Materials and Techniques for Electron Tubes*, Reinhold, New York, 1960.

8. I. Langmuir, *Chem. Revs.*, **13**, 147 (1933).

9. F.J. Norton, *J. Am. Cer. Soc.*, **36**, 90 (1953); also *Trans. 8th Nat. Vac. Symp.*, **8** (1961).

10. R.W. Roberts and T.A. Vanderslice, *Ultrahigh Vacuum and Its Applications*, Prentice-Hall, Englewood Cliffs, N.J. 1963.

11. W.A. Rogers, R.A. Buritz, and D. Alpert, *J. Appl. Phys.*, **25**, 868 (1954).

12. F. Rosebury, *Handbook of Electron Tube and Vacuum Techniques*, Addison-Wesley, Reading, Mass., 1965.

13. C.J. Smithells, *Gases and Metals*, Chapman and Hall, London, 1937.

14. B.J. Todd, *J. Appl. Phys.*, **26**, 1238 (1955); also *J. Appl. Phys.*, **27**, 1209 (1956).

15. P.F. Varadi, *Trans. 7th Nat. Vac. Symp.*, **14** (1960).

Experiment 2.2

COMPARISON OF GAS EVOLUTION PHENOMENA FROM
GLASS AND METAL VACUUM SYSTEM ENVELOPES DURING BAKING

R. P. W. Lawson

Electrical Engineering Department
University of Alberta
Edmonton, Canada

INTRODUCTION

During a vacuum system pump-down cycle, gas is removed from the chamber by

    (a)   volume pumping, and

    (b)   pumping of outgassing species from the envelope

This experiment is concerned with investigation (b), which occurs at low pressures after the volume gas (air initially at atmospheric pressure) has been removed by the roughing pump and the diffusion or ion pump.

Normal outgassing events at low pressures may be grouped under:

    (a)   permeation and diffusion of material through the system walls

    (b)   desorption of material from the inner surfaces of the system walls

These two processes constitute the major limitation to the ultimate pressure achieved within a vacuum chamber--assuming, of course, that there are no leaks present in the system and that the pumps are still effective.

Permeation and diffusion of material through the system walls and into the vacuum is governed by a diffusion coefficient D which is a function of the crystalline nature of the wall material and of the diffusing particle([1]). Normally D is small and it dependent upon temperature T according to:

$$D = D_o \exp\left(-\frac{E}{RT}\right) \tag{1}$$

where E is the activation energy for diffusion, usually expressed in kilocalories per mole; and R is the gas constant.

Any surface of a solid exhibits forces of attraction normal to the surface. Hence gas molecules impinging on the surface are adsorbed. This gas is desorbed under certain conditions of temperature and pressure.

The average time t_s that an adsorbed particle spends on the surface is given by:

$$t_s = \tau_o \exp\left(\frac{E_D}{RT}\right) \tag{2}$$

where τ_o is the period of oscillation of the molecule normal to the surface ($\tau_o = 10^{-13}$ sec), E_D is the activation energy for desorption, and R and T have the usual notation. A full discussion of the significance of Eq. (2) and especially the interpretation of τ_o can be found in the work of de Boer([2]).

Because of the exponential dependence, D and t_s vary over a wide range. Equation (1) indicates that increasing the temperature T increases the diffusion coefficient D; for example, for hydrogen permeating type 300 stainless steel, D lies below 10^{-10} cm^2/sec at room temperature, while at 700°C D rises to 10^{-6} cm^2/sec. Obviously there are limitations to the temperature of baking, but at 200°C D lies within the region of 10^{-8} cm^2/sec.

Helium permeates through Pyrex 7740 glass with a diffusion coefficient D of 10^{-8} cm^2/sec at room temperature, but at 200°C D = 10^{-6} cm^2/sec. The equilibrium pressure within a system is determined by the pumping speed working on the volume and the gas flow into the volume. Since helium diffusion through glass occurs at all temperatures, it will always provide a limiting equilibrium pressure in any particular volume. This limiting pressure is usually acceptably very low at room temperature. Since the diffusion rate increases rapidly with increasing temperature, the diffusion of atmospheric helium through Pyrex, for example, will produce a much higher equilibrium pressure of helium at elevated temperatures. Since helium is chemically inert, this relatively high pressure of helium during bakeout is in almost all practical cases completely tolerable. Thus, when bakeout is completed, the helium is rapidly pumped away to its normally very low equilibrium value.

2.2. COMPARISON OF GAS EVOLUTION PHENOMENON

The mechanism of adsorption depends greatly upon the adsorbent surface and on the adsorbing gas particles which may dissociate into atoms on adsorption or may adsorb as molecules. Weakly bound atoms or molecules will desorb at different times depending upon the temperature of the surface and the desorption energy--E_D in Eq. (2). Extensive treatments of the desorption and permeation processes are beyond the scope of this experiment. Tests are limited to an observation of the release of quantities of gas from the system walls as the temperature is raised.

EXPERIMENTAL SYSTEM

The vacuum system must comprise:

1. Pumping station capable of achieving less than 10^{-8} Torr background

2. Manifold, connecting (a) Pyrex glass bulb (1 liter)
   (b) Stainless steel bulb (1 liter)
   (c) Ionization gauge and control
   (d) Mass spectrometer and control

3. Bakeout mantle for bulbs, with heater supply and temperature monitor (thermocouple)

A schematic of the vacuum chamber is given in Fig. 1.

The manifold can be made of stainless steel #304 and should be baked separately, i.e., apart from the glass and metal bulbs. During the general bakeout (the region within the dotted line) the valves should be open and the bulbs kept at room temperature if possible. When desorbed material from the manifold adsorbs onto the bulbs, their subsequent characteristics will not be materially affected--they will already be contaminated.

The volume of each bulb should be approximately equal in size and equal to the volume of manifold and gauges, so that reasonable mass spectra may be observed during the experiment. Too large a manifold volume may introduce volume effects which "smear out" the mass peaks.

Suggested components:

nude ionization gauge
mass spectrometer
bakeable valves (1 in.)

Either: ion/sorption pumped system
Or: liquid nitrogen-trapped oil-pumped system using Silicon D.C. 705 or Convalex 10 diffusion-pump oil
2 1/2 in. valve in pumping line

EXPERIMENTAL PROCEDURE

It should be possible to carry out the experiment while the manifold and measuring instruments are still hot, but if the background pressure is too high, the system may have to be cooled to room temperature before the experimental period.

Fit the bakeout mantle around one bulb (the other bulb should be sealed off to reduce volume effects). Construction of the mantle can be such that it heats up to ≈400°C in about one hour.

Partially valve-off the pumping line so that the pressure increase is negligible but pumping speed is small. When the pressure is constant, obtain a mass spectrum of the residual gases.

Begin heating the bulb. Monitor total pressure with time and temperature. When significant pressure increases are observed, determine the gas species evolved from the chamber, their maximum

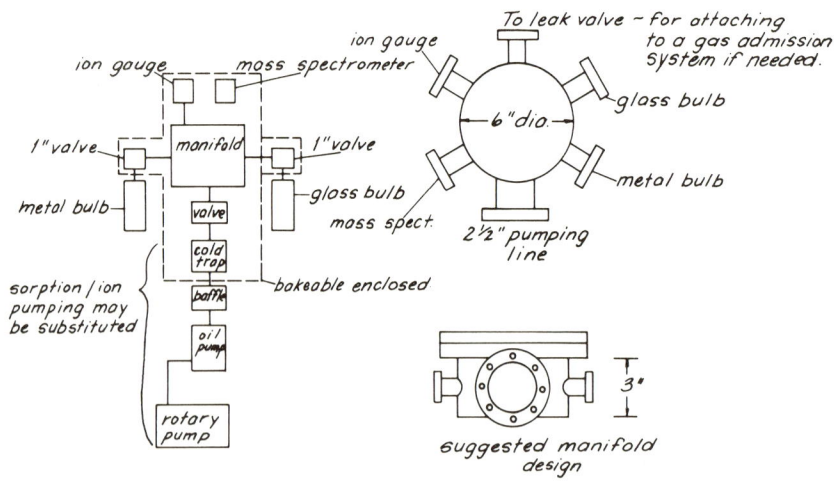

FIG. 1. Vacuum system layout.

2.2. COMPARISON OF GAS EVOLUTION PHENOMENON

release temperature, and the temperature interval of release. When 400°C is reached, keep the temperature constant. Obtain a mass spectrum after the pressure has stabilized. Remove the mantle and allow the bulb to cool, monitoring the total and partial pressures at intervals.

Repeat the experiment for the other bulb, making sure that the pumping line valve is untouched in its partially valved-off position.

Estimates of the outgassing of the species re-evolving from the bulb walls can be made by referring to the American Vacuum Society Standard (3). For this experiment, the throughput method outlined would be adequate, since a known pumping speed of S for the system is more reliable than computing rate-of-rise data in a valved-off system when desorption of many species occurs.

The system pumping speed S must be computed. This may be carried out by admitting gas through a calibrated leak of known conductance K liter/sec. Then

$$K P_r = S P_s \qquad (3)$$

in equilibrium, when P_s is constant, where P_s is the system pressure, and P_r is the gas reservoir pressure.

For different values of P_r Torr, the slope of a graph of P_r versus P_s then yields the pumping speed S. Adequate estimates of pumping speeds S for different gases may be made by applying, for example:

$$S_{H_2O} = S_A \sqrt{\frac{M_A}{M_{H_2O}}} \qquad (4)$$

where the quantity M refers to the mass number and the subscripts A and H_2O refer to the calibrating gas argon and the desorbing gas vapor, respectively.

The ion gauge used to monitor system pressure, and the thermocouple or Pirani gauge used to monitor gas reservoir pressure, must first be calibrated. This may be done in a separate experiment, using either this system with an accurate gauge or by transferring the gauges to a calibration system. The mass spectrometer may be calibrated from the ion gauge.

When monitoring system pressures as the test bulbs are heated, it is advisable to take a mass spectrum at intervals of approximately 50°C in order to obtain a reasonable picture of the gas evolution. Considerable overlap of desorption from the desorbing gases is likely and the ion gauge will only give total pressure readings. The thermal inertia of the heating schedule should be such that the temperature variation with heating time is approximately linear.

A series of graphs of partial pressure versus time (temperature) can then be drawn up for the constituent gases evolved from the test bulb. The integral of the pressure as a function of time over the interval t_2-t_1, during which time the pumping speed S remains constant, gives an indication of the evolved gas quantity Q according to:

$$Q = S/A \int_{t_1}^{t_2} P \, dt \text{ Torr liter/cm}^2 \tag{5}$$

where A is the surface area of the bulb.

The total evolved gas quantity obtained by integrating the ion gauge reading with time may be compared with the sum of the constituent evolved gas quantities obtained with the mass spectrometer. This is a useful exercise for demonstrating the use of a mass spectrometer in analyzing the residual gases present in a high vacuum system.

COMMENTS

It should be possible, from the experimental data, to compare the vacuum outgassing characteristics of glass and stainless steel and to estimate the optimum temperature of bakeout needed to achieve a satisfactory working pressure within a vacuum environment. The mass spectra before and after bakeout may be analyzed; possible reasons for any difference should be attempted.

REFERENCES

1. G. Lewin, Fundamentals of Vacuum Science and Technology, McGraw-Hill, New York, 1965, Chap. 3.

2. J.H. de Boer, The Dynamical Character of Adsorption, Oxford University Press, London, 1968, p. 121.

3. American Vacuum Society Standard, No. AVS 9.1 (1964); J. Vac. Sci. Technol., 2, 6, 314 (1965).

Experiment 2.3

DETERMINATION OF THE NET QUANTITY OF
GAS FLOWING THROUGH A CYLINDRICAL TUBE*

K. M. Busen

Williams College
Williamstown, Massachusetts
and
Sprague Electric Company
North Adams, Massachusetts

INTRODUCTION

The experiment concerns the determination of the quantity of air flowing through a cylindrical tube. A detailed discussion of such a flow is given in Chap. 2 of Ref. 1, where pp. 80-82, 87, 88, and the whole of Sec. 5 are of special importance.

A gas flow can be "viscous," "molecular," or "transitional," depending on whether the gas particles are able to transfer momentum by collisions between themselves, whether they predominantly collide with the walls of the tube, or whether both types of collision occur. There are semiempirical equations for each kind of flow, and it is possible to determine from a parameter L_a/a (with L_a being the mean free path evaluated at the average pressure in the tube and a being the radius of the tube) what kind of flow is prevalent for a tube of given dimensions. For L_a/a <0.01, the flow is viscous; for L_a/a >1.00, the flow is molecular, and for 0.01, <L_a/a <1.00, the flow is transitional.

---

*This experiment is an abbreviated version of one described earlier by the author [Am. J. Phys., 35, 398 (1967)]. The figures are reproduced with the kind permission of the editor of the American Journal of Physics.

The viscous flow is described by the relation:

$$q_v = (B_v/\eta)(a^4/\ell)P_a(P_2 - P_1) \tag{1}$$

where B_v is a constant with a value of 5.236×10^{-4} (g/cm^2/sec^2), η is the viscosity coefficient (g/cm/sec = poise), a is the tube radius (centimeters, ℓ is the tube length (centimeters), P_2 is the pressure at one end of the tube (mTorr), P_1 is the pressure at the other end of the tube (mTorr), and $P_a = 1/2(P_2 + P_1)$ is the average pressure (mTorr). According to the terms used in Eq. (1), the flow has the dimension of millitorr liter per second. The molecular flow follows the relation:

$$q_m = B_m(a^3/\ell)(T/M)^{1/2}(P_2 - P_1) \tag{2}$$

where B_m is a constant with a value of 30.480 [(g cm^2/sec^2)1/mole deg]$^{1/2}$, T is the absolute temperature (degrees Kelvin), and M is the molecular weight (grams per mole). For the transitional flow it can be written:

$$q = \left[\frac{5.236 \times 10^{-4}}{\eta}\frac{a^4}{\ell}P_2 + 30.480Z\frac{a^3}{\ell}\left(\frac{T}{M}\right)^{\frac{1}{2}}\right](P_2 - P_1)$$

where

$$Z = \frac{1 + 2.507\ (aP_a)/L_1}{1 + 3.095\ (aP_a)/L_1} \tag{4}$$

$$Z = 1 \quad \text{for } P_a \ll L_1/a$$

$$Z = 0.81 \quad \text{for } P_a \gg L_1/a$$

and L_1 is the mean free path in centimeters at a pressure of 1 mTorr. As can easily be seen, Eq. (3) reduces to Eq. (2) for $P_a \ll L_1/a$, and practically to Eq. (1) for $P_a \gg L_1/a$. Under conditions of steady-state conservative flow the net quantity of gas across the entrance of a tube is equal to the net quantity of gas at the exit. In this case, the flow is denoted as throughput through the tube. It is common use to measure the throughput in Torr liter per second at 20°C or in mTorr liter per second at 25°C.

The relationships for the throughput through tubes are important for the design of vacuum systems. Often the conductance of a system is reduced unnecessarily by using tubes that have too small a diameter. Similarly, stopcocks with small bores are often an inhibitive factor to effective pumping.

2.3. DETERMINATION OF THE NET QUANTITY OF GAS

EXPERIMENTAL SYSTEM

To determine the net quantity of gas flowing through a cylindrical tube, the pressure is measured at the entrance and the exit of the tube and Eq. (1), (2), or (3) is utilized for computation. In many investigations the pressure at the low-pressure side of the tube is made so small that it may be taken as zero as compared with that at the high-pressure side. "Zero"-pressure can be achieved by connecting the exit of the tube to a pump with a high speed and negligible ultimate pressure. The pressure at the entrance can be recorded by any suitable manometer.

Figure 1 shows the sketch of a vacuum system for the experimental determination of a gas flow. The sketch gives a front view of Parts 1 to 10, whereas Parts 11 to 18 are drawn schematically. In order to obtain a suitable gas flow, a known air-leak is utilized, which delivers a throughput of about 0.5 mTorr liter/sec. The permeation cell characterized by the numbers 1 through 5 and 10 in Fig. 1 is substituted by the air-leak via a joint at B. The joint is connected to a Pirani tube (7) and to a three-way high-vacuum stopcock (8). Connected to the stopcock is a capillary (9) 320 mm long with 2-mm i.d. Other dimensions are indicated in Fig. 1. It is recommended

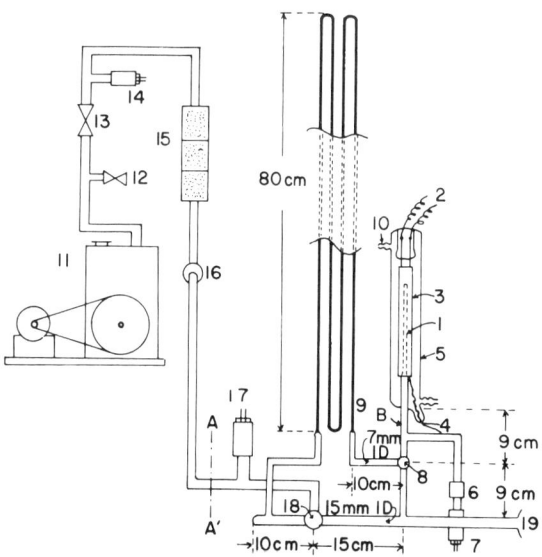

FIG. 1. The vacuum system.

that the glass system to the right of the line AA' be prepared by a professional glassblower. The other parts can be put together by elementary glassblowing techniques. The air coming from the leak is removed through the capillary by means of a pumping system. The components of this system are a fore pump (11), an air-inlet valve (12), a fore-line valve (13), a thermocouple tube (14), a diffusion pump (15), a cold trap (16), and an ion tube (17). The components are specified in Table 1.

At the exit of the capillary, the pumping system has to maintain a pressure which is negligible as compared with the pressure at the entrance. With the set-up present here, there was a pressure ratio of 1:10,000 for all throughputs observed.

A critical part of the system is stopcock (8), which should be of high quality. It is recommended that a stopcock with a hollow plug be used and that it be ground with a lapping compound before installation.

EXPERIMENTAL PROCEDURE

To start the experiment, the pumping system is brought into operating condition with the valves in a position as indicated in Fig. 2(a). For operation of the system, the fore pump is turned on and left running until one reads a pressure of about 100 mTorr at the thermocouple gauge. Then the diffusion pump can be turned on. It takes about 10 min. for this pump to start operating. The system should be left alone now until the ionization gauge indicates a pressure of 0.01 µ or less. This is proof that there are no disturbing leaks.

The ionization gauge is very useful but can be omitted if its expense is unwanted. A further check on the system can be made by having the pumping system pump directly on the open air leak. If the valves are in positions indicated by Fig. 2(c) and the Pirani gauge reads zero, the system is operational and the pumping speed is high enough to keep the pressure sufficiently low at the capillary exit. This test, however, is only valid when pressures of more than 10 mTorr are anticipated at the capillary entrance.

Occasionally one should pump from the other end of the capillary by changing between positions (a) and (b) in Fig. 2, because then the capillary, which has a high impedance, will be more completely evacuated. In the next step the leak is opened. For measurements one must achieve steady-state conditions which exist when the throughput through the capillary is equal to the flow through the air leak. Steady state is indicated when the pressure at the Pirani gauge

2.3 DETERMINATION OF THE NET QUANTITY OF GAS

TABLE 1

Parts List

Number in Fig. 1	Part	Comment
7	Manometer at capillary entrance	Pirani Gauge Tube
6	Vacuum coupling	
8	3-way stopcock, hollow, T-bore 4 mm	
11	Fore pump	
12	Air-inlet valve	1/4-in. globe valve with silicon rubber seat
13	Fore-line valve	High vacuum bellows sealed in-line valve
14	Manometer at fore line	Thermocouple Tube
15	Diffusion pump	Any diffusion pump with a pumping speed of at least 40 liter/sec
17	Manometer at capillary exit (optional)	Ion gauge tube
18	3-way vacuum cup stopcock, 10 mm bore	
	Dust filter	Millipore Filter CS (air) with gasline filter holder
	Flowmeter	
	Pump fluid for diffusion pump	
	Membrane pump	
	Temperature indicating meter	

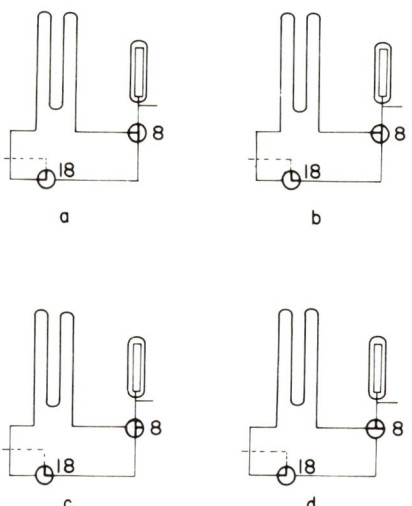

FIG. 2. Operating positions of the valves.

becomes constant (equilibrium pressure). This pressure can be approached either from below or from above. An approach from below takes place when, after outgassing, one changes from position (c) to position (d) in Fig. 2. If it appears that equilibrium pressure is present, one takes a reading. Next, as a good check, approach the pressure from above. For this check, close valve (18) and wait until a substantial pressure increase is observable. Reopening the valve will reduce the pressure to its equilibrium value. Comparison of this value with the previous one gives an idea of how precise the readings are.

When the system is in a steady-state condition, the equilibrium pressure P_2 is read at the Pirani gauge and its value is entered into one of the equations, (1) through (3), depending on the magnitude of L_a/a. The computed gas flow can then be compared to the throughput of the known leak.

Editorial Comment (NM). It is important to realize that conductance and pumping speed are not interchangeable terms. Conductance concerns flow of molecules through ducts, no net work being done. Pumping speed concerns work being done on molecules, thereby establishing concentration gradients. To calculate conductance it is necessary to know (a) the entering conditions, i.e., the angular distribution as a function of velocity and molecular weight of the molecules entering a duct; and (b) the net result in angular distribution only as a consequence of molecular encounters with the duct surfaces. Single and multiple local encounters with the duct wall

2.3. DETERMINATION OF THE NET QUANTITY OF GAS

must be taken into account (5). In the general case, Maxwellian gases do not exist at the entrance(s) of a duct, but when sufficiently Maxwellian gases can be assumed at the duct entrance(s), then do not assume that the molecules enter the duct randomly. Rather it is known (6) that the probability of molecules in a Maxwellian gas entering a duct is porportional to the cosine of the angle between the molecular trajectory and the normal to the entrance plane of the duct. The use of terms such as "random" gas distribution and "random entry are misleading and are to be avoided.

Probability methods offer a dimensionless way of predicting molecular flow through tubes. Under the conditions of steady state, and when Maxwellian gases are entering a tube, a sufficiently large sample of molecules passes its way through the tube and out the other end. Consider two large volumes connected by a tube of arbitrary cross section. In the general case, the cross-sectional shape of the tube will vary from one end to the other. Let the entrance area of such a tube be A_1 and the exit area A_2. If A_1 is not equal to A_2, the chance that a statistically valid number of molecules starting in end 1 will come out end 2 is not the same as the chance that molecules starting in end 2 will come out end 1. The probability, W, of passage through a tube is thus directional and this directionality is denoted by an arrow, i.e., $W_{1\to 2}$. The probability of passage in either direction is defined as the number, N, of molecules entering, divided into the number, M, of molecules exiting. The probability, W, thus has a precision determined by $M^{-1/2}$. This probability is called the Clausing Factor. Note that this Clausing Factor is not an adjustment factor or some sort of boundary-condition fitting constant. The Clausing Factor depends upon the particular gas distribution in space entering a tube, but not on the temperature or velocity of the molecules either initially or subsequently (velocity $\neq 0$).

QUESTIONS

How is conductance related to gas flow probability? Determine for yourself whether or not conductance can be directional. Using the references below, prove that the conductance, $C = V_1/4 \cdot A_1 W_{1\to 2} = V_2/4 \cdot A_2 W_{2\to 1}$; where V_1 is the average velocity of a Maxwellian gas distribution entering the tube at end 1, V_2 is the velocity of a Maxwellian distribution entering the tube at end 2, A_1 is the area of end 1, A_2 is the area of end 2, $W_{1\to 2}$ is the Clausing Factor or probability that a Maxwellian gas entering end 2 will pass out end 1. What can we see from inspection of this relationship? First, that free molecular flow through a tube is determined by the temperature of the gas entering the tube and not by the temperature of the walls of the tube, providing that the wall reflection law is not a

function of wall temperature. Try a test of this
assertion experimentally, being careful not to heat
or cool gas entering the tube and to account for outgassing or pumping effects in the tube. The foregoing
expression also shows the scaling law for free molecular gas flow. All tubes of a geometrically similar
shape have the same free molecular Clausing Factor and
their conductance is then directly proportional to
their entrance area and the velocity of the gas entering. Consider right circular cylinders. If you know
the Clausing Factor as a function of cylinder length
divided by radius, you then can find the conductance
for all right circular cylinders, independent of actual
size.

How do bends in a tube affect free molecular flow?
Test the assertion experimentally that bends in a
tube are of very little consequence under free molecular flow conditions.

How are flow probabilities calculated? In general,
flow probabilities are difficult to calculate. One
calculational method uses a Monte Carlo approach to
simulate a free molecular gas. Refer to the references provided for a discussion of this sort. In a
few cases flow probability can be seen by inspection,
i.e., the probability of flow through a thin orifice
is equal to one. Consider flow through a box that is
large in cross section compared to a pin hole entrance
and a pin hole exit. Is the chance equal to one-half
that molecules entering one pin hole will go out the
other pin hole? Look in one pin hole and shine a
light in the other in order to establish the line of
sight. Now place a small opaque solid in the middle
of the box so that you can no longer see from one pin
hole to the other. Has placing this blocking plate
in the box changed the flow probability through the
box? Replace the small blocking plate in the box by
a venetian blind array stretching from one wall of
the box to the other or by a chevron or Z array
stretching clear across the inside of the box. How
much has the flow probability through the box changed
as a consequence of these internal changes? As the
pin holes are made larger and larger, finally becoming
as large as the whole side of the box, what happens
to the flow probability through the box? What happens
to the gas conductance through the box? If you know
Clausing Factors for individual elements, what is the
Clausing Factor for these elements linked together?

Consider a thought experiment with a pool table.
Arrange your table so that pool balls are put on the

2.3. DETERMINATION OF THE NET QUANTITY OF GAS

table one at a time. Strike the balls with a cue exactly in the center so that no "english" is applied to the ball. If the balls specularly reflect from cushion to cushion without frictional loss, what is the probability that a ball will enter a given pocket? Is the probability of pocketing a ball altered by how hard you hit the ball with the cue? If the balls are deposited on the table at a fixed rate, what is the rate at which they will enter a pocket if the cue stroke has a fixed angle with respect to the table cushions? Does the rate at which balls enter a pocket depend upon the velocity imparted by the cue? Discuss what happens on the table as the number of balls deposited per unit time increases.

What would happen if the wall reflections in a duct were specular instead of diffuse? Discuss the possibilities of making a tube with molecularly smooth walls. Notice the pumping effects reported in Ref. 10.

If gas conductance were not independent of flow direction, could a perpetual motion machine be constructed?

If you roughen the walls of a tube, discuss how the conductance might be affected. Is it possible to redirect molecules by making assymetrical grooves in wall surfaces?

What effect on flow would wall fins or corrugations have?

BIBLIOGRAPHY

1. S. Dushman, Scientific Foundations of Vacuum Technique, (J.M. Lafferty, ed.) 2nd edition, Wiley, New York, 1962.

2. J.O. Ballance, Trans. 3rd Int. Vac. Congress, Stuttgart, 2, 85 (1965).

3. J.N. Chubb, U.K.A.E.A. CLM-R 54 (1965).

4. J.N. Chubb, Vacuum, 16, 591 (1966).

5. D.H. Davis, L.L. Levenson, and N. Milleron, UCRL-6787, (1963), J. Appl. Phys., 35(3), 529 (1964).

6. D.H. Davis, J. Appl. Phys., 31(7), 1169 (1960).

7. L.L. Levenson, N. Milleron, and D.H. Davis, Trans. 7th Nat. Vac. Symp. (1960); and UCRL Report 6014 (1960).

8. C.W. Oatley, Brit. J. Appl. Phys. 8, 15 (1957).

9. P. Clausing, Ann. Phys., 12, 404, 901 (1932).

10. J.P. Hobson, J. Vac. Sci. Technol., 7, 351 (1970).

Section 3

EXPERIMENTS WHICH ILLUSTRATE THE
DEPENDENCE OF THE PHYSICAL PROPERTIES OF GASES ON GAS DENSITY

Experiment 3.1

MEASUREMENT OF THE PUMPING ACTION
OF AN IONIZATION GAUGE

H. Farber

Department of Electrophysics
Polytechnic Institute of Brooklyn
Graduate Center
Farmingdale, New York

INTRODUCTION

An important class of high vacuum and very high vacuum pumps are getter-ion pumps. These are also known as sputter-ion, vac-ion, orbion, or ion-sorption pumps. There are two pumping mechanisms; one is related to the gettering properties of titanium (see Experiment 4.3) while the other is related to the pumping action found in ionization gauges. Magnetic fields, or shaped electrical fields, are used to give the electrons a spiral trajectory to make them more efficient ionizing agents.

Alpert[1] has demonstrated that in a clean, sealed-off section of a well baked vacuum system, ionization gauges may be used to reduce the pressure to the 10^{-9} to 10^{-10} Torr range. There have been many theories [1-4] put forward to explain the pumping action of ionization gauges. In the discussion it is assumed that a clean, small system exists at a maximum pressure of 10^{-7} Torr.

The neutral gas is initially ionized by the electron current flowing between the cathode and grid. Some of these ions are collected by the anode. These ions are the components of the anode current which is related to the pressure measurement. On the other hand, the major fraction of these ionized particles are collected by the negatively charged glass walls of the ion gauge. The glass walls of the gauge are continually bombarded by the charged particles in the tube, which at equilibrium produces a potential on the glass wall. In a clean system the potential of these walls tends to approach the relatively negative potential of the cathode. Observations by Alpert(1), Young(2), and others indicate that three to ten times as many ions are collected on the glass walls than on the anode. These ions are believed to be captured below the surface of the glass and thus are not readily desorbed.

Adsorption or absorption of the neutral gas molecules is generally believed to be relatively very small during the normal operation of the gauge. Thus the major process for removal of the gas molecules initially requires the ionization of the molecules.

In addition to the pumping effect in these gauges, the other dominant effect is the re-emission of the neutralized ions from the glass walls or from the electrode surfaces. Another important effect which becomes noticeable if the pressure is in the 10^{-10} - 10^{-11} Torr range is the diffusion of helium atoms through the glass. This acts to limit the pressure one can achieve in an ordinary glass system (single walled).

Alpert(1) has shown that the maximum pumping speed one could expect from a Bayard-Alpert gauge would be approximately 0.02 liter/sec if the ions collected by the anode were the only pumping mechanism. However, both Alpert and Young observed pumping speeds in the order of 0.1 to 0.2 liter/sec for nitrogen. Young also observed a pumping speed of 4×10^{-3} liter/sec for helium. The following relationship was developed by Young for calculating the rate at which the gases are cleaned up:

$$-\frac{dp}{dt} = a + bp + cp^2$$

where a is largely dependent on the relative rates at which the gas is pumped and the rate at which these particles are re-emitted; and b and c are constants which are dependent on the pumping speed, the relative number of pumping sites, and the probability of sticking when an ion hits an active collecting site.

Robinson and Berz(3) in a later paper suggested the following relationship:

$$p - p_o = c_1 e^{-a_1 t} + c_2 e^{-a_2 t}$$

where the constants are related to the nature of the gas and the surfaces. In addition, the constants a_1 and a_2 are related to the electron current during the pumping operation. They also studied the recovery of the initial pressure when the electron current is switched off. The same relationship is obtained for this condition, except that a_1 is related to the sticking time of the molecules and a_2 is determined by a characteristic of the glass.

3.1. MEASUREMENT OF PUMPING ACTION

An approximation of the pumping speed of an ionization gauge may be readily determined if we assume that only a single pumping mechanism exists in the system; and further, there is no appreciable desorption of gas from any of the surfaces in the vacuum chamber while it is being pumped. For these conditions the pressure at any instant of time may be computed from the relationship:

$$P = P_s + (P_0 - P_s)\, e^{-st/v}$$

or

$$\log \frac{(P_0 - P_s)}{(P - P_s)} = 0.43 \frac{st}{v}$$

where P_s is the steady state or ultimate pressure, P_0 is the initial pressure, s is the pumping speed, and v is the volume of the system.

According to this relationship, a plot of $\log (P_0 - P_s)/(P - P_s)$ against time should be a straight line. As Dushman[4] points out, we are actually measuring the rate of exhaust which is defined as:

$$E = S\left(1 - \frac{P_s}{P}\right)$$

For the condition where $P \gg P_s$, then $E \approx S$ and $P_0 - P_s \approx P_0$, and

$$\log \frac{(P_0 - P_s)}{(P - P_s)} \approx \log \frac{P_0}{P}$$

Since the pumping speed is a function of the nature of the gas and of the electron current, we can obtain a family of curves of $\log P_0/P$ against time for each of these parameters.

EXPERIMENTAL SYSTEM AND PROCEDURE

Pumping speed measurements may be easily made with a clean, baked, sealed-off ionization gauge of the Bayard-Alpert type. (We have noted that ionization gauges that have been used on untrapped oil pumped systems for any extended period of time are not suitable for this experiment, since they tend to give erratic results.)

The tube should be baked out for 4 hours at 400°C while on a properly trapped system. After bakeout, the tube may be flushed with dry nitrogen and the pressure reduced to 10^{-6} to 10^{-7} Torr prior to seal-off.

It is desirable to use an ionization gauge control circuit which permits varying the cathode current at least over the range of 1 to 10 mA. Lower currents are desirable for "zero" readings of initial pressure.

The experiment is carried out by measuring the pressure as a function of time with a fixed cathode current. After the ultimate pressure has been reached, or a change of one decade in pressure has been observed, the cathode current is reduced to its lowest measuring value available and the glass surfaces are gently heated until the higher initial pressure is attained. The tube is then permitted to cool. After cooling, the initial pressure is checked and another set of data is taken at a different cathode current. Momentary heating of the grid will increase the pressure so that measurements may be repeated.

ALTERNATIVE EXPERIMENTS

If a well trapped pumping station is available with a metal bakeable valve, then the experiment can be carried out without sealing off the ion gauge. In this case, after each run the metal valve may be opened; the glass walls may be outgassed by flame-torching, and the next set of data taken after closing the valve. Different gases may be used for purging the tube, such as nitrogen and helium, so a comparison of pumping speeds for different gases could be made.

NOTE: The sealed-off tube is convenient for checking the behavior of the control circuit when the latter requires servicing.

REFERENCES

1. D. Alpert, J. Appl. Phys., 24, 860 (1953).

2. J.R. Young, J. Appl. Phys., 27, 926 (1956).

3. N.W. Robinson and F. Berz, Vacuum, 9, No. 1 48 (1959).

4. S. Dushman, Scientific Foundations of Vacuum Technique, (J.M. Lafferty, ed.), 2nd edition, Wiley, New York 1962.

3.1. MEASUREMENT OF PUMPING ACTION

5. P.A. Redhead, J.P. Hobson, and E.V. Kornelson, <u>The Physical Basis of Ultrahigh Vacuum</u>, Chapman and Hall, London, 1968.

<u>Editorial Comment</u> (NM). An ionization gauge not only can pump but can become a source of gas as well. Reference 5 gives a thorough discussion of pumping and outgassing effects. What tests can you contrive to check whether or not a gauge is a source or a sink for a particular component of the gas phase?

Experiment 3.2

STUDY OF THE LINEARITY
OF AN IONIZATION GAUGE

J. R. Miller, III

U.S. Army Metrology and Calibration Center
Redstone Arsenal, Alabama

INTRODUCTION

The linearity of ion gauges can be investigated by establishing a pressure rise which is linear with time. A method based upon Knudsen's laws of gas flow was described as long ago as 1921 by Dushman and Found [1] to check the linearity of ion gauges over several decades of pressure.

The following equations are derived and applied to the apparatus in Fig. 1:

$$Q = \frac{d}{dt}(PV), \text{ and } Q = \Delta PC \qquad (1)$$

$$V_1 \frac{dP_1}{dt} = (P_a - P_1)C_1$$

$$dP_1 = \frac{C_1}{V_1}(P_a - P_1)\,dt$$

where $P_a \gg P_1$

hence

$$\int_{P_{10}}^{P_1} dP = \frac{C_1}{V_1} P_a \int_0^t dt$$

FIG. 1. Vacuum gauge calibration system.

$$P_1 - P_{10} = \frac{C_1}{V_1} \cdot P_a t^n \tag{2}$$

A linear increase of the ion current (igk = constant) with time constitutes a check on the linearity of the gauge. If we take the logarithim of Eq. (2) we get

$$\ln (P_1 - P_{10}) = n \ln t + \ln \frac{C_1 P_a}{V_1}$$

which is the equation of a straight line $[\ln(P_1 - P_{10})]$ plotted versus $\ln t$ with the slope n equal to the exponent of t, which should equal 1.

EXPERIMENTAL SYSTEM AND PROCEDURE

The ion gauge used should be attached to the volume V_1 using ultrahigh-vacuum techniques; glass fusing to a glass to metal seal, copper gaskets, bakeable valves, etc. The system is prepared by baking at 350-400°C for an extended time at high vacuum to remove gases. P_a is monitored with a mechanical gauge such as the Wallace and Tiernan at around 1 atm. Nitrogen or helium may be used.

The rate of rise of pressure in V_1 may be monitored in one of two ways. Intermittent readings can be made by turning the ion gauge on and off periodically to prevent pumping. The on-time should be short compared to the off-time. The electron current should also be reduced to 100 μA or below. This method requires a certain damping in the electrometer circuit to prevent shutting off the circuit. This method also responds to the very rapid desorption of gas that is adsorbed on the filament and one sees the pressure go through a very short-lived transient--reach a plateau and begin a slower reduction in pressure due to pumping action. Read the plateau.

3.2. LINEARITY OF AN IONIZATION GAUGE

Another method may be used, although at very low pressures it may require a special electrometer. Reduce the emission current to 100 μA or below and operate the ion gauge continuously. The very low electron current reduces pumping.

To start the procedure, valve off from the pumps at some low pressure, say, 10^{-8} Torr, and slowly open C_1 (which may be a variable leak) to get an increasing pressure as read with the ion gauge. Start a timer and proceed to take data.

Plot your results on log-log paper and calculate the slope.

This experiment will cover 3 or 4 decades in a reasonable time if the original leak if large enough.

COMMENTS

It is instructive to check the linearity of Bayard-Alpert gauges and other types of ion gauges both for helium and nitrogen in the high pressure area, say, 10^{-5} to 10^{-2} Torr to observe some of the nonlinearities(_2_).

Discuss how pumping and outgassing of an ion gauge may affect linearity(_3_). (See Experiment 3.1.)

REFERENCES

1. S. Dushman, <u>Scientific Foundations of Vacuum Technique</u> (J.M. Lafferty, ed.) 2nd edition, Wiley, New York, 1962.

2. W.B. Nottingham and F.L. Torney, <u>Trans. 7th Nat. Vac. Symp.</u>, 117 (1960).

3. P.A. Redhead, J.P. Hobson, and E.V. Kornelson, <u>The Physical Basis of Ultrahigh Vacuum</u>, Chapman and Hall, London, 1968.

Experiment 3.3

CALIBRATION OF GAUGES

C. F. Morrison

Granville-Phillips Company
Boulder, Colorado

INTRODUCTION

Presently accepted methods for wide-range vacuum gauge calibration tend to require very expensive equipment. In addition, they perform a seriously limited function, in that the available low pressure transducers tend to be very sensitive to gas composition. Unless the gas composition is known in the measurement situation, even the pressure decade may remain in doubt, in spite of the use of a calibrated gauge. As a further difficulty, gas flow in vacuum systems can create unexpected pressure drops, and cause anomalous gauge behavior. This somber introduction is not meant to discourage, but to forewarn the scientist or engineer hoping to make quantitative or semiquantitative use of low pressure technology. If experiments in which submillimeter pressures are variables are to be meaningful, careful attention must be given to the design of the entire system, and its use, as well as to the calibration of the gauges involved.

Providing wide-range vacuum gauge calibration experience without indulging in a 20-100 thousand dollar calibration system will require a number of compromises. These will be decreased range and increased time and care required to achieve meaningful accuracy. Even with expensive calibration equipment, certain of the operations prove to be very time consuming. The experiments given here do not necessarily fit nicely into standard laboratory periods, but will require prior planning and system preparation.

NATURE OF THE PROBLEM

Pressures within the range of the oil or mercury manometer may be measured with accuracy and direct traceability to the fundamental standards. Calibration of vacuum transducers in this pressure range can thus be the operation of direct comparison with a suitable manometer. At pressures below about 10^{-2} Torr the difficulties increase seriously. It seems an elementary extension of the manometer to trap a known, large volume of the low pressure gas, compress it to a known smaller volume, and there measure it with a manometer. The compression manometer, or McLeod gauge, has indeed served for many years as an unofficial standard in the pressure range below 10^{-1} Torr. Because this gauge uses mercury as the compressant and manometric agent, it is necessary to trap the mercury vapor such that it does not contaminate the system being measured. The vaporization-trapping cycle is all too similar to that in a mercury diffusion pump. The resulting pressure errors, which are both pressure and composition dependent, have largely led to the abandonment of the McLeod gauge for standard work. Other gauge types do not tend to be absolute in this lower pressure range, either, so one is forced to generate calibration pressures which are referenced to higher pressures, or to other measurements, such as that of gas flow.

Two basic methods are used for establishing knowable calibration pressures from higher, directly measureable pressures: volume expansion and gas flow through series conductances. Volume expansion is based upon Boyle's law. A small known volume of gas at sufficiently high pressure for accurate measurement is expanded into a larger, previously evacuated, known volume in order to generate a calibration pressure. Because the calibration chamber is not actively pumped during the calibration proper, this is often referred to as a "static" calibration method. In such a method, any gas source or sink will create an error. The method is therefore of major significance for calibrating gauges which do not pump, i.e., which do not create ions in the operation of their gauging mechanisms, and for calibrating at pressures where significant sorption and outgassing do not take place. To be sure, there are differences of opinion as to the practical use of such systems for other than the rare gases, and as to the allowable pressure range. The volume expansion method is included in this experiment in that it makes possible reasonably certain calibration in the range of pressures not easily readable on the liquid manometer--nor accurately generated with conductance methods.

So-called "dynamic" pressure generation--as opposed to "static"--is that in which the system is actively pumped during calibration, and the pressures are generated by gas flow. In the molecular flow regime, the flow of gas through a limiting conductance generates a pressure difference (ΔP Torr) across the conductance which is simply related to the conductance (C liter/sec) and the gas flow (Q Torr liter/sec).

3.3. CALIBRATION OF GAUGES

$$\Delta P = Q/C = P_1 - P_2 \tag{1}$$

Ideally, P_2, the pressure below the conductance, would be sufficiently near zero that it could be ignored, making the value of P_1, the pressure above the conductance, readily calculable. However, the speed of the pumping system (and the series valve in the experimental system) are such that P_2 is a significant fraction of P_1. If the gauge giving the P_2 reading was calibrated, the results would again be simple to interpret. Simultaneous calibration of this gauge and the others is accomplished in the experiment.

Use of two conductances in series permits the generation of a pressure ratio. Such a ratio has found much use in extending calibration pressures to very low values. Such a pressure ratio can be developed between the reference system and the volume above the limiting conductance of the experimental system. Procedures are given for using these ratios to generate calibration points.

Pinning down calibration values based upon pressure ratios and known conductances requires either the measurement of an absolute pressure, or a measurement of a gas flow rate. The reference system of the calibrator contains a pressure measuring system which operates in the range of 1 to 100 Torr. This part of the system is a manometer, or can be compared with a manometer for its calibration. The reference system also contains a known fixed volume. Use of this volume in conjunction with the pressure meter makes possible the metering of flow rates.

EXPERIMENTAL SYSTEM

Figure 1 shows the schematic of the calibration system. The system consists of a liquid nitrogen trapped-oil diffusion pump with suitable roughing. Above this very conventional pumping system is a valve of considerable conductance which can be used to change the system from "static" to "dynamic," and which can be used to isolate the pump when changing gauge tubes on the calibration chamber.

The calibration chamber is a 15-cm diameter tube with three to eight gauge port flanges. The center of this chamber should contain a thin plate orifice 0.0125 cm thick (5 mil), with a 0.4688-cm diameter hole bored through its center. This orifice serves as the limiting conductance referred to in the previous section. The orifice is made by clamping the 5-mil stock between heavier pieces of metal, then precision boring through all three parts. The orifice plate is either clamped or welded into the center plate of the chamber.

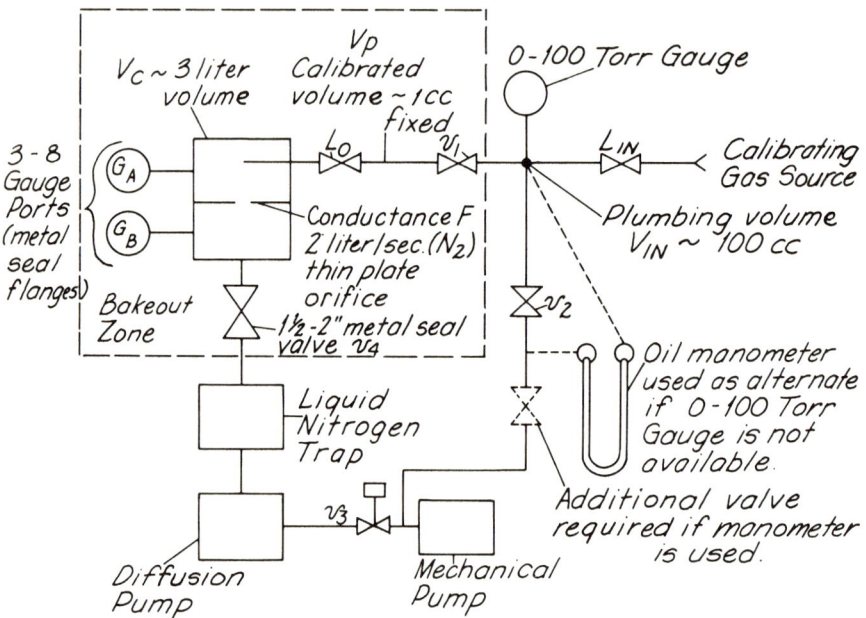

FIG. 1. Vacuum gauge calibration system.

The gas inlet to the calibration chamber contains a vacuum "pipet," or calibrated volume of about 1 cc. The plumbing volume of the inlet system should be about 100 cc. If an oil manometer is used in place of the suggested 0-100 Torr meter, considerably more care is required in the operation of the system. Why?

The roughing pump has been provided with a solenoid valve such that this same pump can be used to remove large quantities of gas from the reference system without interrupting the diffusion pump operation.

3.3. CALIBRATION OF GAUGES

EXPERIMENTAL PROCEDURES

A. STATIC BACKFILLING

1. Preliminary System Measurements

   Determination of V_{in}. The volume of the inlet system, V_{in}, is of importance in determining the calibration chamber volume, in forming the largest and smallest pressure steps, and in practicing the inlet volume method of dynamic calibration. The method used for this determination will be to fill the calibration chamber to atmospheric pressure, evacuate the inlet volume, then pipet gas from the calibration chamber into the inlet until a sufficiently high pressure exists there to measure accurately with the 0-100 Torr meter.

   Calculate the value of V_{in} from the first pipet of gas released into the inlet system:

   $$V_{in} = \frac{P_{atm} V_p}{P_{in}} - V_p$$

   where P_{atm} is the initial pressure in V_p prior to expansion into V_{in}, and P_{in} is the final pressure in V_{in}.

   How can the results of multiple transfers of gas into the system be calculated accurately?

   Determination of V_c. The approximate volume of the calibration chamber, V_c, can be calculated from the dimensions of the system. However, for most accurate calibration it is desirable to measure this volume. For this procedure it is necessary that V_{in} be known. In this procedure, the inlet is filled to atmospheric pressure, and this gas is expanded into the calibration chamber to give a pressure readable on the 0-100 Torr meter.

   Calculate

   $$V_c = \frac{(V_{in} + V_p)(P_{atm} - P_2)}{P_2}$$

   where P_{atm} is the initial pressure in the inlet volume V_{in} plus V_p prior to expansion of this quantity of gas into V_c by opening valve L_o, and P_2 is the final pressure in the total volume ($V_c + V_{in} + V_p$).

2. **Calibration with Argon**

Evacuate V_c and V_p to the ultimate pressure that can be obtained using the diffusion pumping system.

   a. Close valves V_3 and seal leak L_o

   b. Open V_2 and V_1 to pump out inlet system

   c. Admit argon through L_{in} such as to purge the inlet system and reference volume

   d. Close V_2 and leak L_{in}

   e. Choose initial pressure step size to be at least equal to the indicated background pressure and at least equal to the practical calibration interval suggested by the graduations on the gauge dial

   f. Determine the procedure by which this step size can be accomplished. Based upon V_c = 3000 cc, V_p = 1 cc, V_{in} = 100 cc, and a 0-100 Torr meter with accuracy from 3 to 100 Torr, the following steps are easily achieved:

   3×10^{-4} Torr/step -- Fill inlet to 100 Torr. Close V_1, close V_3, open V_2. After inlet is thoroughly pumped out close V_2, open V_3, open V_1. Gas trapped in pipet expands to fill inlet volume. This should provide an accurately calculable inlet pressure of ∼1 Torr--below the accurate range of the meter.

   1×10^{-3} Torr/step -- Fill inlet to 3 Torr

   3×10^{-3} Torr/step -- Fill inlet to 10 Torr

   1×10^{-2} Torr/step -- Fill inlet to 30 Torr

   3×10^{-2} Torr/step -- Fill inlet to 100 Torr

   2×10^{-1} Torr/step -- Fill to atmospheric pressure

   1 Torr/step   -- Fill to P_c + 31 Torr (dump entire inlet into calibration chamber)

   g. Fill pipet by pressurizing the inlet system as per f, then closing V_1

3.3. CALIBRATION OF GAUGES

    h. Record the gauge readings

    i. Open leak valve L_o to admit the calibrating gas to the chamber

    j. Record the gauge readings

    k. Close L_o and open V_1 to refill the pipet

    l. Close V_1

    m. Repeat h through l until the step size is no longer adequate, then choose a new step size and repeat g on, until the range of the transducer has been covered.

    n. Calculate calibration pressure values from the known values of V_c, V_p, V_{in}, and the reading of the inlet system meter. Are corrections needed in the filling pressure value as successive steps are used from the inlet?

B. DYNAMIC CALIBRATION

1. Pipet Method

Principle. By knowing the pipet volume V_p, pressure P_{in}, and temperature, an exactly known quantity of gas is trapped in the pipet. The known quantity of gas may be used in conjunction with the known conductance of the orifice, F, to generate a constant pressure, P_r, for as long as the gas will last. This time, t_p, along with the conductance and gas quantity can be used to calculate the absolute pressure of the chamber during the determination:

$$P_r = \frac{P_{in} \cdot V_p}{F \cdot t_p} + P_b$$

In this equation, P_r is the chamber pressure in Torr during the determination; P_{in} is the pressure in Torr at which the pipet was filled; V_p is the volume of the pipet in liters ($\sim 1 \times 10^{-3}$ L); F is the conductance of the lower orifice in liter/sec, calculated for the gas in use, $F = 2.00 \times (28/M)^{1/2}$ L/sec (M = molecular weight of calibrating gas); and t_p is the time, in seconds, required to use up the gas. The effective value of t_p may be taken from the pressure versus time recording as shown in Fig. 2. P_b is the pressure below the orifice.

The volume V_p shown in Fig. 1, bracketed between the leak valve, L_o, and valve V_1 has been very carefully measured. This known volume

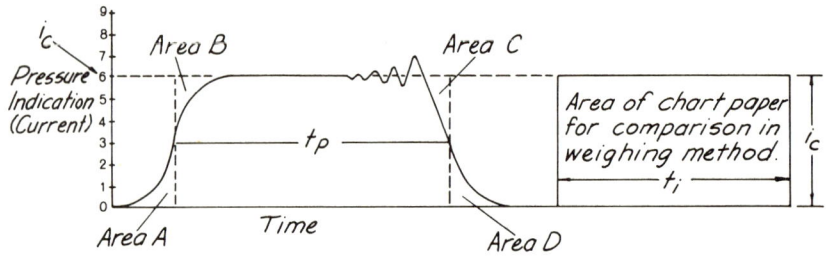

FIG. 2. Pressure indication-time curve for pipet.

is filled to a known pressure, as read on the inlet system meter. Valve V_1 is closed, trapping the known quantity of gas, $P_{in}V_p$, in the pipet. The pressure indication (current)-time curve is then recorded while the leak valve is used to maintain the system pressure-indication constant at a convenient calibration value. The area under the chamber pressure indication-time curve is then used with the volume of the pipet, V_p, and its filling pressure, P_{in}, to calculate the chamber pressure in Torr corresponding to the measured pressure indication (current). Because the rise to the control point, and the fall from it at the end of the run, are not instantaneous, it is necessary to correct the time value for the starting and stopping characteristics. These most often can be satisfactorily allowed for by measuring the time from the point at which pressure reached one-half the required control point, to the point at which the pressure fell back to this same value at the end of the run. (See Fig. 2.)

The estimate of effective time at reference pressure can be precise to several percent--1% with considerable care and numerous repetitions to improve the statistics. However, the 1% figure can often be attained on a single run by calculating the ion current ampere-seconds of response for the pipet of gas. Use of this technique implies linearity of the gauge in that pressure decade, so it should be applied with some caution. This method is the quantitative application of the concept illustrated in Fig. 2 when areas A and B and C and D are seen to be very nearly equal. Instead of assuming that areas are symmetrical about the starting and stopping times on the pipet run, the number of ampere-seconds of response are determined by counting enclosed squares of the recorder paper under the pressure signal-time curve.

If the pressure indication-time curve is sufficiently irregular to be difficult to interpret in the manner presented, consider the following technique. Carefully cut out with scissors the area of recorder chart representing the pressure indication-time curve. Cut out a similar area, this time perfectly rectangular, representing a constant pressure indication, P_i, of the desired value, and an exact

3.3. CALIBRATION OF GAUGES

time, t_i, of the same order as the experimental time. Weigh these two pieces of chart paper on an analytical balance.

$$\frac{\text{(Weight of curve)}}{\text{(Weight of rectangle)}} \times t_i = t_p$$

where t_p is the time required to empty the pipet maintaining the calibration pressure at P_i. This method also requires the assumption that the pressure indication is linear with pressure over the range of the recorder excursions, and the assumption that the chart paper is uniformly dense. Both conditions are usually met sufficiently to permit the order of 1% repeatability.

In the basic equation for calibration pressure, the pump pressure, P_b, is a necessary correction. This is, at best, the indication of a calibrated gauge. In the absence of such a gauge, minimal error is usually introduced by using:

$$P_b = K P_r$$

where K is determined by comparing response of the gauge for P_b with another gauge (with both on the same chamber), transferring the second gauge to chamber A and determining the pressure ratio.

2. Inlet Volume Method

Principle. The volume, V_{in}, is constant, fixed by the plumbing of the system. The pressure in this volume is read on the 0-100 Torr meter. At constant, known temperature, gas is used from this volume, V_{in}, such that the pressure there changes from an initial value P_1 to a value of P_2. The quantity of gas, G, used is, thus:

$$G = (P_1 - P_2)V_{in}$$

A known quantity of gas flowing at constant rate in a known time, t_r, through the known conductance, F, into the pump generates a known pressure, P_r, above the conductance:

$$P_r = \frac{G}{t_r F}$$

Measuring the time, t_r, required to use the gas out of the inlet system such as to change the pressure there from P_1 to P_2 while maintaining a calibration pressure, P_r, in the calibrator, provides sufficient information to establish the value of that calibration pressure:

$$P_r = \frac{(P_1 - P_2)V_{in}}{t_r F}$$

Some caution is advised, in that the lower the fraction of the gas used from the inlet system during the measurement, the greater is the possible error due to changing temperature. If about one-half of the gas is used (P_2 = 5 Torr, P_1 = 10 Torr), the effect of a 2°C change in the system temperature is about 1% in the calibration value. If only 1% of the gas is used for the measurement (P_1 = 100 Torr, P_2 = 99 Torr), a 1°C change can cause a 30% error. Therefore, it is recommended that approximately one-third to one-half of the inlet system gas be used in the measurement, i.e., P_1-P_2 should be about one-half of P_1.

3. Pressure Ratio Generation

Depending upon the nature of the leak valve L_o, and the absence of impurities, especially water, in the inlet system, it may be possible to establish proportional behavior between the inlet system pressure and the calibration pressure. When this proportionality is shown to exist, it is then possible to vary the inlet pressure to vary directly the calibration pressure. This makes practical the taking of many points in the calibration curve. Also, this makes possible a stepping operation which can extend the range of the system to lower pressures, limited only by the system ultimate pressure capability.

To establish the proportionality necessary for this type of operation:

    a. Fill the inlet system to 10 Torr

    b. Open the leak L_o to establish a calibration chamber pressure about 100 times the ultimate pressure-- record this value.

    c. Increase the inlet system pressure to 20 Torr

    d. The calibration chamber pressure should double

    e. If the calibration pressure doubled, increase the inlet pressure to 40 Torr and check again

    f. If the pressure did not double, return the inlet pressure to 10 Torr to check the original reading. If the system does not return to the previous pressure indication, moisture in the leak valve may be the problem. However, it may well be the case that all leaks will not exhibit the required molecular flow characteristics, and this method can not be used with them.

If the leak is shown to be proportional, the following sequence can provide calibration over a wide range:

3.3. CALIBRATION OF GAUGES

1. Pump system to ultimate pressure
2. Pressurize inlet to 10 Torr
3. Open leak valve to give first $P_{cal} = 10\ P_{ult}$
4. Record all gauge readings
5. Increase inlet pressure to 20 Torr
6. Record all gauge readings
7. Decrease inlet pressure to 10 Torr
8. Open leak to re-establish gauge readings from step 6
9. Return to step 5
10. When pressure has been stepped-up in this manner a sufficient number of times that the calibration pressure can be determined by the pipet method, use it to establish the value of the last pressure generated in the calibration chamber.
11. From this known pressure value, calculate the previous calibration points using the inlet-pressure proportionally.

4. Gauge Pumping Determination

If the gauge above the orifice operates as a pump, the calibration pressure will be less than that calculated. The orifice (for nitrogen) represents 2.00 liter/sec. If the gauge pumps to the extent of 0.1 liter/sec, this represents about 5% error. Obviously, if two gauges are operated above the orifice, a 10% error is probable. Unfortunately, the gauge pumping is not always pressure-independent or constant--so it should be determined for accurate work.

   a. Install two gauges above the orifice
   b. Degas gauges at low pressure
   c. Operate at appropriate calibration pressures P_1
   d. Change from 10 mA emission to 1.0 mA on one gauge while measuring pressure P_2 with the other gauge
   e. If the gauges are clean, the pressure will increase due to a 90% decrease in pumping speed of the one gauge

f. Calculate the pump speed of the gauge:

$$S_{10mA} = \frac{(P_2 - P_1)S_{orifice}}{(P_1 - 0.1\ P_2)} \text{ liter/sec}$$

g. How can this information be used to correct the earlier data?

The foregoing exercises represent several of the major techniques available for vacuum gauge calibration. Obviously, similar operations can be performed with less operator work and measurement detail using expensive calibration equipment. However, with sufficient care these methods may be applied with significantly precise results. There are still arguments among the authorities as to the absolute accuracy of the orifice formulas: How did the dynamic method results in your experiments compare with the static measurements? What are the sources of error in these experiments?

BIBLIOGRAPHY

General Reviews:

C. Morrison, Trans. Inter. Vac. Metal. Conf., 101 (1968).

C. Morrison, Research/Development, (September 1969).

Specific Methods:

R.S. Barton and J.N. Chubb, Vacuum, 15, 113 (1965).

C.E. Normand, Trans. 8th Nat. Vac. Symp., 534 (1962).

N.A. Florescu, Trans. 8th Nat. Vac. Symp., 504 (1962).

EDITORIAL COMMENT (NM)

1. Suppose calibration of a vacuum gauge is performed on a calibration system; however, the calibrated gauge is then used on another vacuum system. When the calibrated gauge has been transferred to the system where it is to be used, does the calibration for the gauge still hold? Discuss possible sources of error.

3.3. CALIBRATION OF GAUGES

2. New ionization gauges made by the same manufacturer in the same batch may exhibit a 20% variation in positive current output from one gauge to another.

3. Discuss the readings obtained from ionization gauges as a measure of vacuum environmental conditions. Discuss the feasibility and desirability of in situ calibration.

4. Discuss the relevance to vacuum problems of calibrating gauges in terms of the time rate of transfer of momentum in a gas.

5. The United States National Bureau of Standards declines to calibrate ionization manometers. From a technical standpoint, discuss possible reasons why. Hint: are readings traceable to "fundamental" constants of length, mass, time, and temperature?

6. Is "vacuum pressure" a colloquial notion?

Section 4

EXPERIMENTS WHICH EXAMINE PHYSICAL
AND CHEMICAL INTERACTIONS AT SURFACES

Experiment 4.1

STUDY OF THE SORPTION OF GASES FOR
DIFFERENT GAS-SORBENT COMBINATIONS

K. B. Wear

1114 Fayetteville Rd. S.E.
Atlanta, Georgia

INTRODUCTION

    A gas or vapor molecule chancing into the fields of force near a solid or liquid surface must participate in an interchange of energies with the surface molecules. The very coalescenece of molecules into the liquid or solid is evidence of fields of force surrounding the molecules; these fields are generally attributed to electronic shell structure and valence electrons.

    Further, there seems no possible arrangement of coalesced atoms or molecules that wholly satisfies their fields both in the bulk material and in that abrupt region generally recognized as "surface." As the forces between molecules, in and at the surface, give rise to such localized structure as crystallographic arrangement or such extensive phenomena as surface tension, so the imbalance of forces at and above the surface give rise to residence sites for stray molecules and attractive fields in the neighborhood of the surface. Thus, a gas molecule chancing within the surface fields will yield energy to or derive energy from the molecules comprising the surface. Sorption, which is the tendency for accumulation of gas molecules at the surface, is evidently one facet of the definition of a "surface" and the study of energetics in, at, and near the surface.

    A number of highly specific gas-surface molecular pairs produce a distinct chemical by-product or reaction, e.g., phosphorus pentoxide plus water vapor, or crystal arrangements as in hydration reactions. These reactions are excluded from studies of sorption. A second group of highly specific gas-surface molecular pairs produce valence bonds through the sharing of electrons by host (sorbent) and guest (sorbate) molecules; thus chemical adsorption, or chemisorption, behaves as a chemical reaction but produces no distinct by-product. All solid and liquid surfaces, even though inert in the sense that the valency requirements of their atoms are satisfied, exhibit a physical attraction for any gas, even the inert or noble gases; thus physical adsorption, or physisorption, is much like the attraction and condensation of a vapor on the surface of its own liquid. Because physisorption and chemisorption cannot, in all cases, be clearly distinguished, both are taken together as sorption. (Desorption, a process similar to evaporation and the inverse of chemisorption and physisorption, is also included in the term sorption, although desorption is commonly studied by other techniques.) Movement of chemisorbed or physisorbed molecules into the interior of the solid or liquid, while not experimentally distinguishable from sorption, is not described as gas-surface phenomena and is thus excluded from the term sorption.

Sorption takes place with a decrease in surface free energy. Further, inasmuch as gases lose degrees of freedom upon assuming positions on the surface, entropy decreases. Thus enthalpy must also decrease and sorption is always exothermic.

Sorption of most sorbent-sorbate pairs may be described by:

$$Q = f(N,T) \tag{1}$$

with concentration N and temperature T expressed as absolute quantities. Quantity Q or sorbate may be expressed, per unit mass of sorbent, in moles or in volume at STP. Inasmuch as data is often taken with the sorbent temperature controlled at a fixed value as with a constant temperature bath, the isotherm has been widely used to present sorption data. Typical isotherms are shown in Figs. 1 and 2, which are data by Magnus and Kraft[1] as included in Dushman's[2] classic book.

A family of isotherms may be readily replotted as sorption isobars (pressure constant). Temperatures must be interpolated from the family of isotherms, and are the abscissae of the family of isobars. Data may be replotted as isosteres (constant Q), each of which is analogous to the vapor-pressure versus temperature curve of a liquid. As with a vapor-pressure curve, a form of heat of sorption q_s is given by the Clausius-Clapeyron equation:

$$q_s = - R_o \frac{d\ln p}{d(1/T)} \tag{2}$$

where R_o is the gas constant per mole.

4.1. SORPTION OF GASES

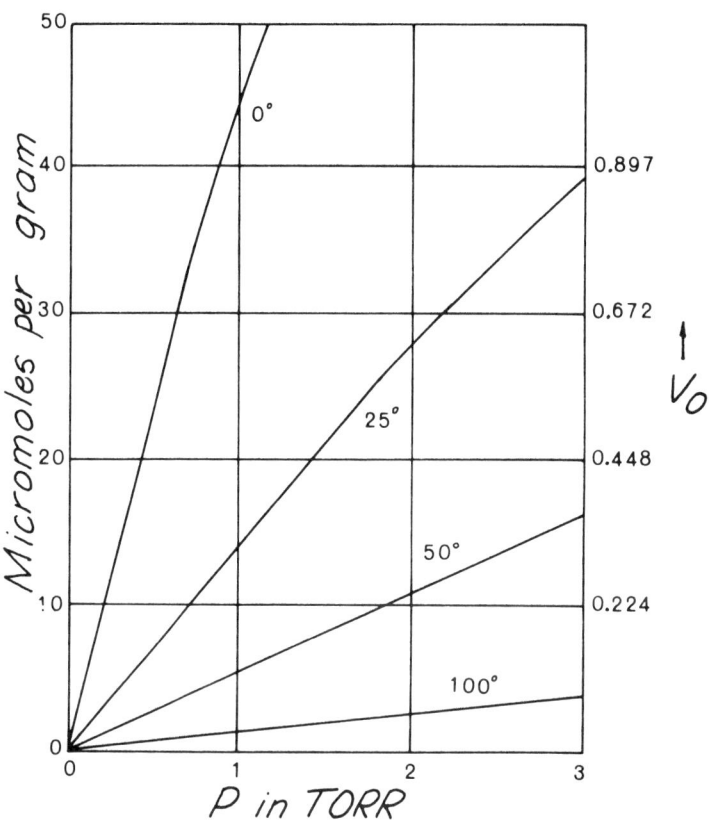

FIG. 1. Isotherms for the sorption at low pressures of carbon dioxide by wood charcoal. After A. Magnus and H. Kraft, Z. Anorg. Chem., 184, 241 (1929).

EXPERIMENTAL SYSTEM

Adsorption isotherms may be easily generated with the apparatus shown in Fig. 3. A stopcock leading to an immersible finger (perhaps 20 mm in diameter by 30 mm tall, mounted on an entrance tubulation) is substituted for the mixing volume. The immersible finger may be partially filled with a known quantity of a very few grams (perhaps between 2 and 5) of sorbent such as titanium dioxide or synthetic zeolite.

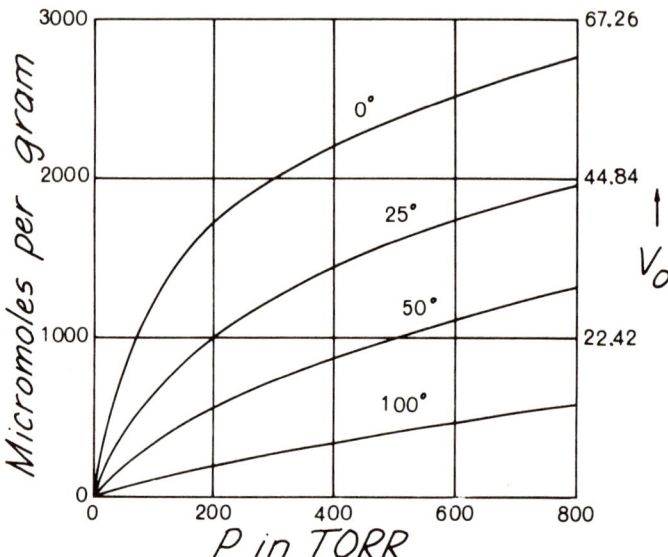

FIG. 2. Isotherms for the sorption at higher pressures of carbon dioxide by wood charcoal. After A. Magnus and H. Kratz, Z. Anorg. Chem., 184, 241 (1929).

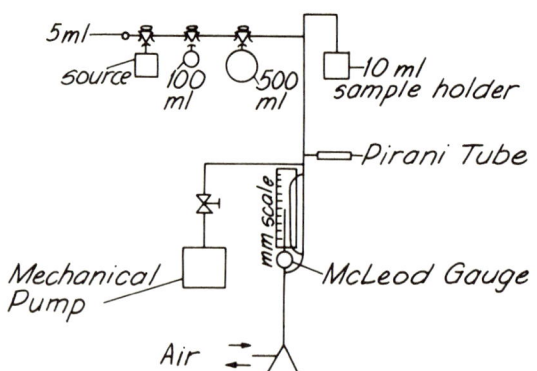

FIG. 3. Experimental system.

4.1. SORPTION OF GASES

The sorbent may be activated by heating with an electric heater or torch during a latter stage of evacuation, being careful to heat gently at first to preclude bursts of gas that can entrain and redistribute sorbent, yet heating virorously enough to thoroughly divest the sorbent of sorbed gases. The finger may be maintained during sorption at a constant temperature such as steam, room, ice, evaporating refrigerant, dry ice, liquid nitrogen, or the melting point of a selected liquid frozen by dry ice or liquid nitrogen. Quantities of gas admitted may be metered and/or measured with the gas inlet arm shown or a more elaborate arm, or with the McLeod or thermal conductivity gauge (if calibrated against the McLeod gauge for the gases being used), and the quantity sorbed (in Torr-liters) may be calculated as the quantity admitted less the quantity necessary to increase the pressure within the system volume. (Glassware volumes may be measured if one volume is accurately known, since gas captured in the known volume at a pressure measured with the McLeod gauge may, after the volume to be measured and the McLeod gauge have been evacuated, be expanded into the volume and gauge and the new pressure be used to determine the volume.)

EXPERIMENTAL PROCEDURE

After charging the immersible finger and evacuating the system (including the volume above the liquid in the source flask), the sorbent may be activated as outlined above.

Before proceeding, system pressure and condensable content should be measured with the McLeod gauge, since the sorbent may require further activation.

1. At room temperature, allow a known quantity of ethanol (ethanol vapor pressure at room temperature times volume of metering bulb) to flow to the sorbent, titanium dioxide. After a few minutes (thermal conductivity gauge constant), measure the quantity of ethanol sorbed and the pressure, by measuring the ethanol pressure remaining over the sorbent.

2. Repeat the addition of ethanol, each time measuring the quantity sorbed and the resultant pressure, until the resultant pressure approaches 10% or so of the equilibrium vapor pressure. Q versus P may be plotted as an isotherm. (In calculating total ethanol available for sorption, account for loss to evacuation of McLeod inlet if this has been done to improve accuracy as when pressure reaches 1/4-1/2 Torr and thus exerts a discernible pressure not accounted for in the McLeod equation.)

3. Re-evacuate and re-activate the sorbent. With the immersible finger at the ice point, repeat the step-wise additions of ethanol. Q versus P yields a second isotherm.

4. Additional isotherms may be determined at other temperatures so long as ethanol vapor is not condensed due to temperature alone.

5. Isotherm data may be readily replotted as T versus P at constant Q (isostere) or as $\ln P$ versus $1/T$ at constant Q, the slope of which yields q_s.

ADDITIONAL EXPERIMENTS

The following additional experiments can use the same experimental system:

1. Nitrogen substituted for ethanol, and temperatures extended as far as liquid nitrogen.

2. Ethanol in nitrogen as a two-gas sorbate. (So long as the percentage of ethanol in the mix is high, the McLeod gauge may be used to measure the partial pressures of both condensable and permanent gases.)

3. Sorbent of synthetic zeolite of 5-Å pore size.

4. Water vapor sorbate.

BIBLIOGRAPHY

1. A. Magnus and H. Kratz, Z. Anorg. Chem., 184, 241 (1929).

2. S. Dushman, Scientific Foundations of Vacuum Technique, (J.M. Lafferty, ed.) 2nd edition, Wiley, New York, 1962.

3. P.A. Redhead, J.P. Hobson, and E.V. Kornelsen, The Physical Basis of Ultrahigh Vacuum, Chapman and Hall, London, 1968.

Experiment 4.2

THE USE OF SORBENTS AS TRAPS AND PUMPS

H. Farber

Department of Electrophysics
Polytechnic Institute of
Brooklyn Graduate Center
Farmingdale, New York

INTRODUCTION

The sorbent properties of materials such as charcoal and molecular sieves (see Experiment 4.1) are the basis of several useful devices in vacuum systems. Three such devices, which are studied in this experiment are:

1. an oil trap to prevent fore pump oil vapors from contaminating a system

2. an oil trap to prevent diffusion pump oil from reaching very high vacuum chambers

3. a sorption pump to eliminate the need of a mechanical fore pump

Although the devices used in part 1 and in part 2 are traps, they operate in different gas flow regimes. This means that the design of the trap must be different. The trap in part 1 operates in the pressure range of 10^{-4} to 10^{-5} Torr. Since the sorption pump operates in the same region, the basic design of the two may be the same except that provisions must be made for cooling these with liquid nitrogen (1,2). When the unit is used as a sorption pump, no mechanical pump is necessary, and the sorbent material is cooled to liquid nitrogen temperatures.

The earlier sorbent traps were usually filled with activated charcoal, and this is still being used by many workers. Recently, since 1950, artificial zeolites (molecular sieves) have been developed with prescribed adsorption characteristics. The artificial zeolites are metal aluminum silicates where the metal cation may be sodium, potassium, or calcium. Depending on the specific cation used and the basic crystallite structure, pore sizes ranging from 3 to 10 Å may be achieved. The pore size then determines what molecules may be adsorbed by this material. Because of the porous structure of these zeolites, the useful adsorbing surface is approximately 600 m^2/g of material.

A very desirable feature of these materials is their regenerative capabilities. By heating these materials to 400-500°C even in air, almost all of the adsorbed gas is released. Further, at these temperatures any coking of organic materials will tend to be removed by oxidation. In normal use a heating mantle around the trap may be used to heat up the trap while it is still being pumped, in order to regenerate the material. When the molecular sieves are used in sorbent pumps, permitting the trap temperature to rise to ambient temperature will sufficiently regenerate the material so that the pump may be used several times before a heating cycle is required to restore the original pumping speed.

Another sorbent material that has been suggested is "thirsty glass"(3) (available from Corning Glass). This material is formed in an intermediate step in the preparation of Vycor glass. This material avoids the problem of molecular sieves, which tend to powder. This powder may be troublesome if it is accidently blown into any metal valves in the system.

EXPERIMENTAL SYSTEM

The molecular sieve traps and pumps that are used in this experiment may be purchased commercially or they may be constructed quite readily using some of the many designs in the literature(1-6). Most of the commercial traps have heating elements, so the molecular sieve material may be regenerated by baking the material at 300-400°C for several hours (some authors recommend as much as 20 hours) while at pressures of 10^{-3} to 10^{-4} Torr. A trap designed by Biondi is shown in Fig. 1. Figures 2, 3, and 4 show some of the commercial traps that may be purchased.

4.2. SORBENTS AS TRAPS AND PUMPS

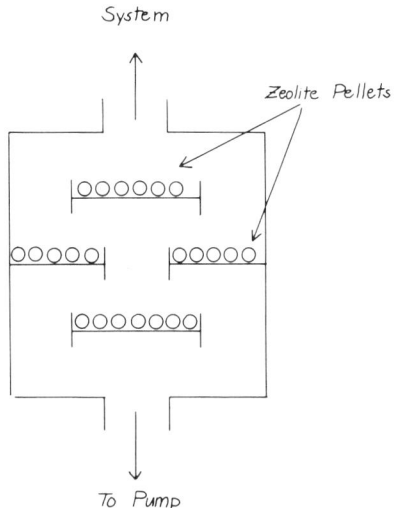

FIG. 1. Biondi-type trap. Reprinted from <u>Transactions of the National Vacuum Symposium</u> with the permission of the editor.

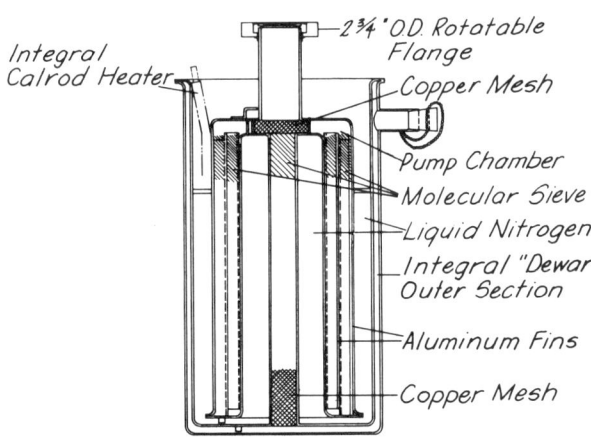

FIG. 2. Cross section of Molecular Sieve Sorption pump. (General Electric Company Catalog GEZ-415.)

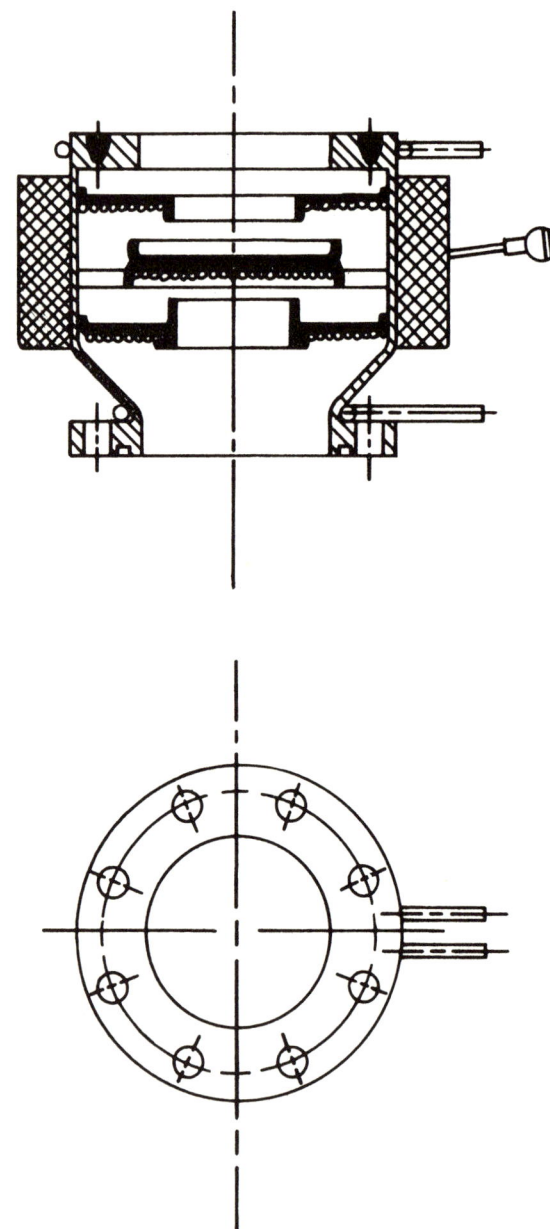

FIG. 3. Molecular sorbent baffle. (NRC Series HZ, Type 0317. National Research Corporation)

4.2. SORBENTS AS TRAPS AND PUMPS

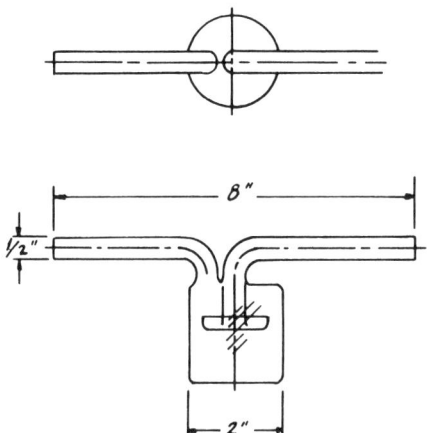

FIG. 4. Pyrex glass sorbent trap. (Consolidated Vacuum Corporation CVC Type TSG-52)

EXPERIMENTAL PROCEDURE

The effectiveness of a molecular sieve trap may be demonstrated using the arrangement shown in Fig. 5. If the trap can also be cooled with liquid nitrogen, then the same set-up can be used to demonstrate sorption pumping.

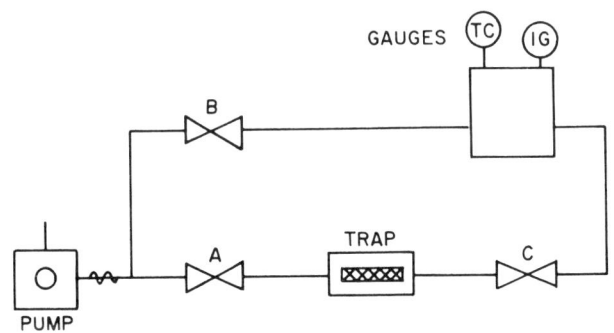

FIG. 5. Set-up for studying molecular sieves.

Initially, the trap is thoroughly baked at 200-400°C, with valves B and C closed, for 4-6 hours; or the sorbent material may be baked in a separate chamber and added to the trap prior to use(6). In carrying out the experiment, valve B is opened and valves A and C are closed. Then the pressure is monitored as a function of time. When a steady-state pressure is reached, usually 1.5×10^{-3} Torr, valve B is closed and valves A and C are opened, and pumping is continued, and the pressure as a function of time is recorded. Finally valve A is closed and liquid nitrogen is added to the sorbent trap and the pressure is again monitored. A typical curve is shown in Fig. 6.

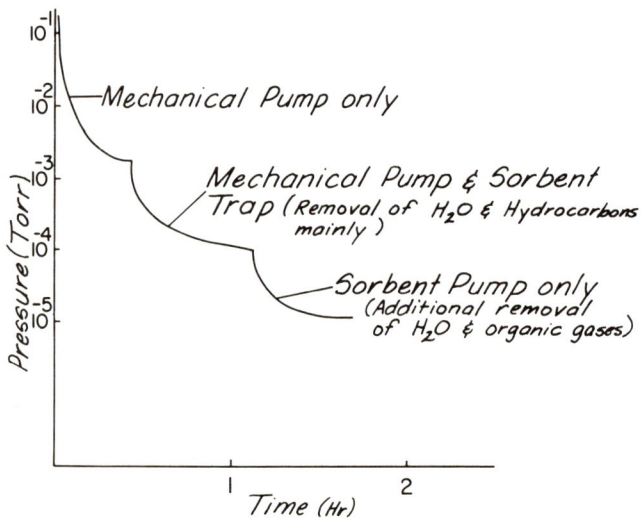

FIG. 6. Typical pumping curve for three-stage operation.

As an alternate procedure, each step is carried out after the vacuum chamber is let up to atmospheric pressure. Pumping speed calculations may be made using the same procedures that are outlined in the experiment on ionization gauge pumps (Experiment 3.1). Since the volume is not known accurately, the computations will only be indicative of the relative pumping speeds.

The advantages of using a molecular sieve trap(4), [or a thirsty glass trap(3)] in the high vacuum range may be demonstrated with the experimental set-up shown in Fig. 7. A small oven which can be placed over the ionization gauge and the trap, as shown by Alpert(8), is also required. Without baking the tube or the trap, the ultimate pressure

4.2. SORBENTS AS TRAPS AND PUMPS

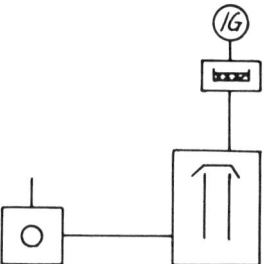

FIG. 7. Experimental arrangement with sieve trap between diffusion pump and ionization gauge.

that can be reached with the diffusion pump is noted. This pressure is usually 10^{-6} to 10^{-7} Torr, depending on how clean the system is initially. Then the baking oven is lowered over the trap and the ionization gauge, for a bakeout at 400°C for four hours. After cooling, the pressure is again noted. The pressure should now be approximately 10^{-8} Torr or lower. This low pressure may be maintained for extended periods of time while connected to the diffusion pump before the trap shows appreciable deterioration in performance(4).

ALTERNATIVE EXPERIMENTS

For the experiments demonstrating sorption pumping, the system may be filled with different gases and pumping speed measurements may be determined for each gas. A comparison could be made for such gases as air and helium.

REFERENCES

1. T.H. Batzer and R.H. McFarland, Rev. Sci. Inst., 36, 328 (1965).

2. P.L. Read, Vacuum, 13, 271 (1963).

3. F.B. Haller, Rev. Sci. Inst., 35, 1356 (1964).

4. M.A. Biondi, Trans. 7th Nat. Vac. Symp., 24 (1960).

5. D. Goerz, Trans. 7th Nat. Vac. Symp., 65 (1960).

6. W.W. Roepke and K.C. Pung, Vacuum, 457 (1968).

7. P.F. Varadi and K. Ettre, Trans. 7th Nat. Vac. Symp., 248 (1960).

8. D. Alpert, J. Appl. Phys., 24, 860 (1953).

Editorial Comment (NM). The following precautionary statements regarding zeolite and other molecular sponge materials should be noted. Materials with very high specific areas per unit weight are necessarily easily broken up into very fine fragments. Accordingly, a very fine dust or powder is always associated with the use of these materials. Unless extreme care is taken, this dust from the molecular sponge material distributes itself throughout a vacuum system, getting into sensitive areas. The dust represents an extremely large reservoir of gas and so the apparatus must be either baked out or cooled down to eliminate substantial outgassing from the dust. Heat transfer in vacuo to and from the dust is poor. Also, if a molecular sponge dust is a dielectric, serious electrical charging effects can develop near plasmas, ion beams, electron beams, etc. Caution: particles of dust may be caught in an all metal valve upon closure, causing the valve to leak seriously.

It should also be noted that zeolites are not effective as sorbent baffles for oil vapor removal when exposed to water vapor.

Experiment 4.3

SORPTION OF GASES BY TITANIUM

H. Farber

Department of Electrophysics
Polytechnic Institute of Brooklyn,
Graduate Center
Farmingdale, New York

INTRODUCTION

In Experiment 4.2 the sorption of gases by materials is studied. In this experiment an application of the gettering action of titanium is examined.

Getters have been used in sealed-off electronic tubes from the early days of tube production. By sorbing large quantities of gas, the getter will markedly reduce the pressure in the electronic tube after it is sealed off, and it will maintain high vacuum during the life of the tube. Getter materials that have been used include such metals as aluminum, magnesium, barium, thorium, and zirconium. Barium is one of the most common metals used today. [Dushman[1] has an extensive survey of gas sorption by metals.]

Another very important application of getters is in sputter-ion or getter-ion pumps. In these pumps titanium metal is continuously sputtered or evaporated during normal operation. The fresh film of titanium rapidly takes up large quantities of gas. In addition, as the titanium is deposited on the surfaces of the pump, it occludes some of the gases which it does not sorb. These two processes make possible the high pumping speeds of the sputter-ion pumps.

Many authors have studied the gettering action of titanium[2,3]. They have shown that titanium is an effective getter for a large number of gases which include oxygen, nitrogen, carbon dioxide, carbon monoxide, water, and hydrogen. Many gases such as oxygen and nitrogen are best sorbed when the titanium is operated at elevated temperatures such as 700°C(4). At this temperature the compounds formed on the surface are dissolved in the metal, permitting additional surface absorption.

Hydrogen will be desorbed when the titanium is heated, and in the temperature range of 400-900°C large quantities of hydrogen may be desorbed from a strip of titanium which has been loaded with hydrogen. Since hydrogen is the only gas released in any appreciable quantity at these elevated temperatures, a foil loaded with gas may be an excellent source of pure hydrogen. One commercial application of this technique is in large hydrogen thyratrons such as tube type numbers 5C22 and 5948.

EXPERIMENTAL SYSTEM

The pressure-temperature relationship of hydrogen gas above hot titanium may be used to demonstrate the gettering action of titanium. Since this is a reversible phenomenon for this metal-gas system, a sealed-off tube may be the most convenient way for carrying out the experiment.

The experimental chamber consists of a glass chamber, 1-1/2 liters in volume, equipped with a thermocouple gauge and an ionization gauge (or a single high pressure ionization gauge), and contains a hydrogen source recovered from an old hydrogen thyratron. These sources, which consist of a cylinder made of titanium foil which is concentric and integral with a heater, are very convenient. The unit can be mounted on a standard tube stem which is then sealed into the system.

Provided that the titanium is not exposed to air while hot, it will be suitable for this experiment. If an oxide film forms on the titanium, it will not absorb appreciable quantities of hydrogen.

The vacuum station which is described in detail below can be used for filling sealed-off gas discharge tubes containing spectroscopically pure helium or hydrogen. It can be used also for student experiments which demonstrate: (a) the effectiveness of molecular sieve traps; (b) ion pumping using a Bayard-Alpert ionization gauge; (c) diffusion rates of gases through metals or quartz; and (d) gettering action of titanium for hydrogen and its inability to getter helium; as well as the processing techniques which are also described in some detail below.

The station which is shown schematially in Fig. 1 may be divided into two sections. The pumping section consists of a

4.3. SORPTION OF GASES BY TITANIUM

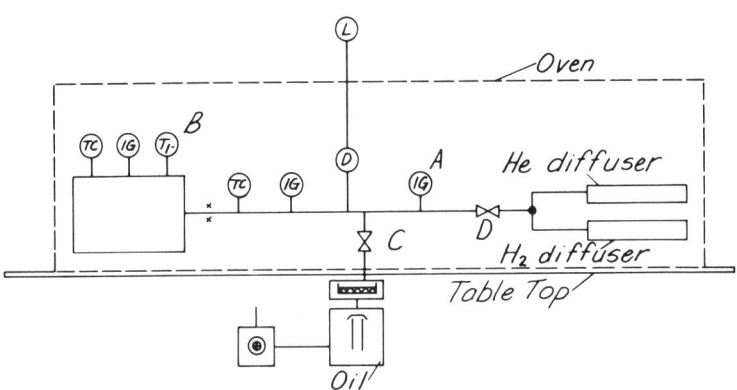

FIG. 1. Vacuum system. A. Ionization gauge, B. Ti-hydrogen source mounted on tube stem, C. and D. Metal vacuum valves, x-x. Seal-off constriction to permit removal of vacuum chamber after processing.

two-stage mechanical pump, an air-cooled 50 liter/sec oil-diffusion pump and a metal molecular sieve trap similar to the one described by Biondi(5). Flanges which use copper gaskets are used to connect the diffusion pump to the trap, and the trap to a kovar-to-glass transition. The remainder of the station is mounted on a marinite (similar to transite) table top which also serves as a base for an oven. The two metal valves have kovar-to-glass seals, so that all of the components shown in the oven section, Fig. 1, are assembled by the glass blower. The helium diffuser and the hydrogen diffuser are shown schematically in Fig. 2 and their characteristics are given in Figs. 3, 4, and 5. The characteristics are for the General Electric Company Permeation Leak Gas Purifier and are given in the General Electric Company Catalog GEZ-4019. The thermocouple gauge is calibrated with an oil manometer using the bakeable capacitance manometer as a pressure null detector. By this arrangement, that gauge can be calibrated after bakeout with no oil contamination.

The oven is an insulated, metal rectangular box, open at the bottom, with strip heaters mounted on the sides. It is light weight and can easily be lifted into place. Suitable ovens are described by Dushman(1) and Alpert(4). This oven sits on the table during bakeout of the station.

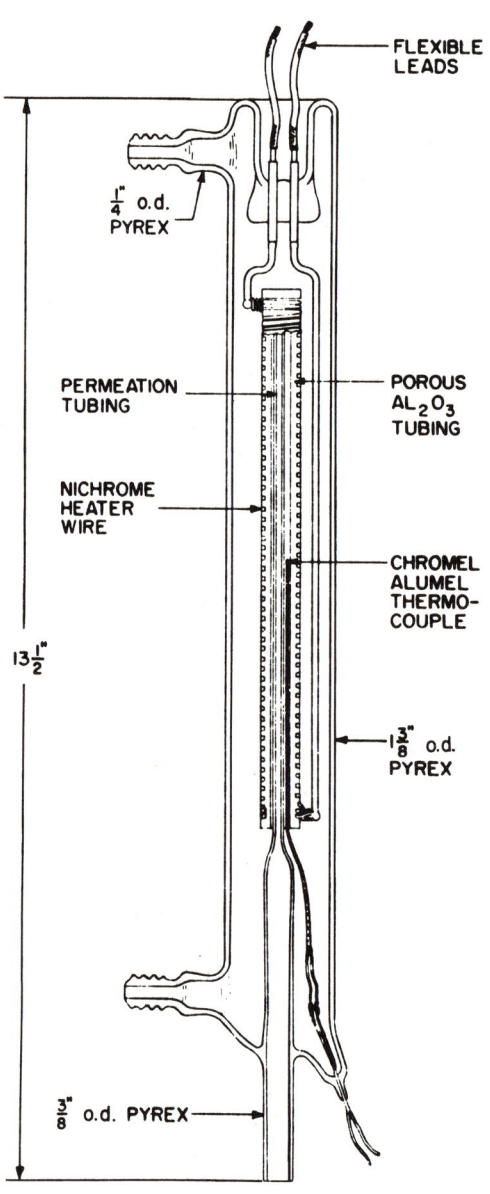

FIG. 2. Schematic diagram of a typical gas diffuser. (General Electric Company Catalog GEZ-4019.)

4.3. SORPTION OF GASES BY TITANIUM

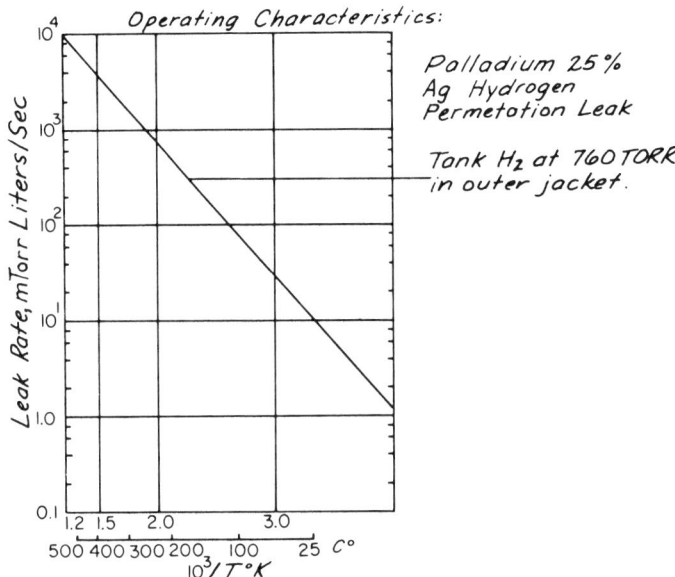

FIG. 3. Operating characteristics of palladium 25% Ag-hydrogen permeation leak. (General Electric Company Catalog GEZ-4019.)

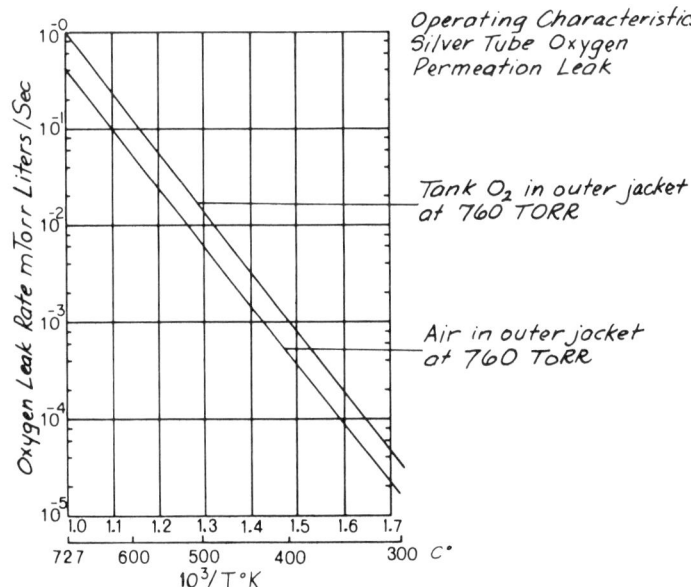

FIG. 4. Operating characteristics of silver tube oxygen permeation leak. (General Electric Company Catalog GEZ-4019.)

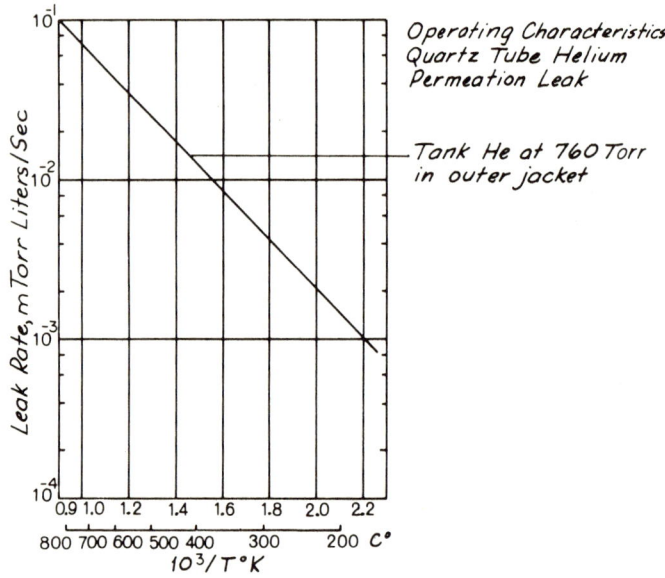

FIG. 5. Operating characteristics of quartz tube helium permeation leak. (General Electric Company Catalog GEZ-4019.)

EXPERIMENTAL PROCEDURE

The system can usually reach a pressure of 10^{-6} - 10^{-7} Torr within an hour after turning on the diffusion pump without any additional processing. At this point the system can be "flamed" gently and the diffusers heated to 300-400°C. Alternatively the oven can be used to quickly raise the temperature to 400°C.

Valve A is closed and the molecular sieve trap is baked at 300-400°C for four hours (or overnight, depending on convenience). After cool-down valve A is opened. The oven is lowered over the station and a 4-hour (or overnight) bakeout at 400°C is used. When the station has cooled, the pressure should be in the 10^{-8} - 10^{-9} Torr range. The helium quartz diffuser is raised to 800°C, the hydrogen diffuser is raised to 400°C, and the titanium hydrogen source is brought to 10% above its normal operating voltage (usually 3.5-5 V). (In our experiments we have always used units which were recovered from old hydrogen thyratrons. The assembled hydrogen sources also may be available from the tube manufacturer.)

4.3. SORPTION OF GASES BY TITANIUM

After these units have been outgassed and have subsequently cooled down to room temperature, valve A is again closed. There may be a small burst of gas, but the ionization gauge now acts as a vacuum pump(4) and the pressure will decrease slowly to a new ultimate low value. The temperature of the diffusers and the titanium foil should not exceed the magnitude used for outgassing as long as valve A is closed.

Gas Filling Procedure

The titanium may now be "loaded" with hydrogen.

Tank hydrogen is now permitted to flow through the outer chamber of the diffuser. With the set-up shown in Fig. 6, the system should be flushed quickly with hydrogen and then a very slight positive pressure of hydrogen is maintained in the diffuser as evidenced by occasional bubbles in the second flask. The temperature of the diffuser is slowly raised to approximately 100°C. The leak rate for the unit is about 0.1 Torr liter/sec at this temperature. The preferred diffusion rate for any purpose may be easily adjusted when using the hydrogen diffuser. Diffusion of helium through quartz is not as convenient, since the rate of diffusion is so low.

The pressure in the system is now monitored by the calibrated thermocouple gauge. The diffuser is permitted to cool as soon as the hydrogen pressure exceeds the maximum operating pressure by 25% (usually 2-4 Torr). The titanium source is then permitted to cool and the pressure in the chamber should be 10^{-8}-10^{-9} Torr.

Before sealing off the discharge chamber, valve A is opened. Thus, most of the outgassing vapors will be pumped out and not trapped in the discharge chamber.

FIG. 6. Suggested system for filling with hydrogen.

Normally valve B is not needed. However, it is very convenient if helium gas is used. Since the diffusion rate of helium is so low, the diffusion can be either carried out overnight and the helium stored at an elevated pressure by closing valve B; or, it may be done during the day while some other experiments are being carried out in the discharge chamber. Further, if proper care is taken while opening valve B, then the valve can be used to control the gas pressure in the discharge chamber. In this way, experiments may be carried out over a wide range of pressures without long delays in building up a suitable pressure of helium.

Notes

A. The application of gaseous diffusion to obtain high purity gases has been reported by many authors. Whetten and Young (6-8) have shown that higher purity gases can be obtained by using tank gas and a diffuser than by using spectroscopically pure "liter flask" gases. In many cases it is also more convenient. Different materials have been used for different gases, e.g., nickel, palladium (1), and palladium-silver (25%) (6) for hydrogen, quartz (7) for helium and silver for oxygen (8). In some electrodeless discharge studies, traces of nickel have been observed spectroscopically when using a nickel diffuser.

B. Titanium source: Instead of the indirectly heated cylinder of titanium described above, directly heated foils of titanium may be used. Morrison (2) describes a unit 7 in. long x 1/16 in. wide and 6 mils thick. Stout (3) describes his experiments using a ring of titanium which was indirectly heated by an induced RF current.

C. Spectrometer studies: In our laboratory the discharge tube is used to study high power pulsed, electrodeless hydrogen discharges. If a monochrometer or a visible light spectrometer is available, then the spectra of the hydrogen discharge may be studied. Even after several months of operation, only hydrogen lines were seen in the spectra. The addition of two electrodes to the sealed-off chamber would provide a convenient hydrogen discharge tube for spectra studies as well as the experiments on getters.

ALTERNATIVE EXPERIMENTS

Several additional experiments that may be performed on this vacuum station are outlined in the introduction.

The diffusion rate of helium and hydrogen through the quartz tube as functions of temperature (1) may be compared. It will be observed that the diffusion rate

4.3. SORPTION OF GASES BY TITANIUM

of hydrogen is approximately 10% of the diffusion rate of helium. Similar experiments can be done with the palladium-silver diffuser.

Another experiment would be to compare the current-voltage relationship for the titanium heater for the conditions of high vacuum and for a hydrogen atmosphere. The latter curve may be used to give an indication of the hydrogen pressure. In carrying out this experiment the time constant for establishing equilibrium conditions as the applied voltage is changed should be noted.

REFERENCES

1. S. Dushman, Scientific Foundations of Vacuum Technique (J.M. Lafferty, ed.), 2nd edition, Wiley, New York, 1962.

2. J. Morrison, Trans. 6th Nat. Vac. Symp., 291 (1959).

3. V.L. Stout and M.D. Givvons, J. Appl. Phys., 26, 1488 (1955).

4. D. Alpert, J. Appl. Phys., 24, 860 (1953).

5. M.A. Biondi, Rev. Sci. Instr., 30(9), 831 (1959).

6. N.R. Whetten and J.R. Young, Rev. Sci. Instr., 31, 1113 (1960).

7. N.R. Whetten and J.R. Young, Rev. Sci. Instr., 32, 453 (1961).

8. N.R. Whetten and J.R. Young, Rev. Sci. Instr., 30, 472 (1959).

Editorial Comment (DL). Another method for working with the hydrogen-titanium system is to use titanium hydride powder. This powder is readily available commercially. It can be placed in a simple cup and held there by spot welding a screen over the open area. The cup can be fabricated to be heated directly or indirectly. Since in this form the material has the maximum initial quantity of hydrogen in the titanium, a small charge will provide an excellent and copious source of hydrogen when the powder is heated. [Reference: L. Levine and D. Lichtman, Rev. Sci. Instr., 31(7), 731 (1960).]

Reply to Editorial Comment (HF). It should be noted that if high purity hydrogen were required, the system would normally be baked at 400°C. Further, the hydride would have to be partially outgassed to remove any absorbed gases other than hydrogen which it absorbed while exposed to air. As a result, some hydrogen would be pumped out and the hydride would require a regeneration process similar to the one given in the experiment. (Although I was aware of the powdered titanium hydride source, I have never worked with it and so the above is only a conjecture. Since I required high purity hydrogen, I preferred forming the hydride in the discharge chamber.)

Experiment 4.4

INVESTIGATION OF THE PASSAGE OF
OXYGEN ACROSS A SILVER BARRIER[*]

K. M. Busen

Williams College,
Williamstown, Massachussetts
and
Sprague Electric Company,
North Adams, Massachussetts

INTRODUCTION

It has been known for a long time that gases are capable of passing from one side of a solid to the other by a process called "permeation." Normally this process is the total of several partial processes which, to a greater or lesser extent (depending on the forces between the gas molecules themselves and between the gas molecules and the atoms of the solid), can take place at the boundaries of the solid and its interior. A detailed discussion on these partial processes is given in Ref. 1.

Permeation is measured in terms of a gas flow which follows the relation:

$$Q = Q_o \exp[-E_{perm}/(RT)] \tag{1}$$

where E_{perm}, the activation energy, is the energy necessary to bring the gas from one side of the solid barrier to the other, and Q_o is a constant. By combining the two gas flows Q_1 and Q_2, which are measured at temperatures T_1 and T_2, one obtains by means of Eq. (1):

---

[*]This experiment is an abbreviated version of one described earlier by the author [Am. J. Phys., 35, 398 (1967)]. The figures are reproduced with the kind permission of the editor of the American Journal of Physics.

$$Q_2/Q_1 = \exp[-E_{perm}(T_2 - T_1)/RT_1T_2] \qquad (2)$$

Often permeation is determined as the total of two partial processes, namely absorption and diffusion. If the diffusion coefficient D does not depend on concentration and if the solubility s of the gas in the solid is given by Henry's law, $s = \sigma P$, (where σ is the solubility constant and P is the gas pressure), then in the stationary state:

$$Q = D\sigma \Delta P/d \qquad (3)$$

In the last equation $\Delta P = P_2 - P_1$ is the pressure difference for the two sides of the solids and d is the thickness of the barrier. When, at a given temperature ΔP and d are taken as unity, the value of the product $D\sigma$ in Eq. (3) is defined as permeability, π. In the above, the solubility s was taken as being proportional to the pressure P. Often it is found to be proportional to $P^{1/2}$. Choosing $P_1 = 0$, one obtains from Eq. (3):

$$Q = \pi P^{1/2}/d = \text{const } P^{1/2} \qquad (4)$$

This linear relation between the flow and the square root of the pressure is found for gases which exist as molecules outside the barrier, but after dissociation diffuse as free atoms through the solid.

An example for permeation is offered in the next section, where it is shown how the activation energy for the permeation of silver by oxygen can be determined experimentally and in what state oxygen diffuses through silver.

EXPERIMENTAL SYSTEM

Measurement of permeation of a solid barrier is simple because it requires the determination of a gas flow only. The measuring procedure is explained in the text.

The barrier used in this experiment is a "permeation lead-gas purifier," oxygen leak. It consists (see Fig. 1) of a silver tube (1) 6 in. long with an outer diameter of 150 mil and a wall thickness of 10 mil. The tube can be heated by connecting a variable ac power supply to the terminals (2) of the heating element (3). The oxygen flow is varied by changing the temperature of the heating element, and the temperature is monitored by an indicating meter connected to the terminals (4) of the chromel-alumel thermocouple which is located between the silver tube and the heating element. The Pyrex jacket (5) around the heating element allows for the passage of

4.4. PASSAGE OF OXYGEN ACROSS A SILVER BARRIER

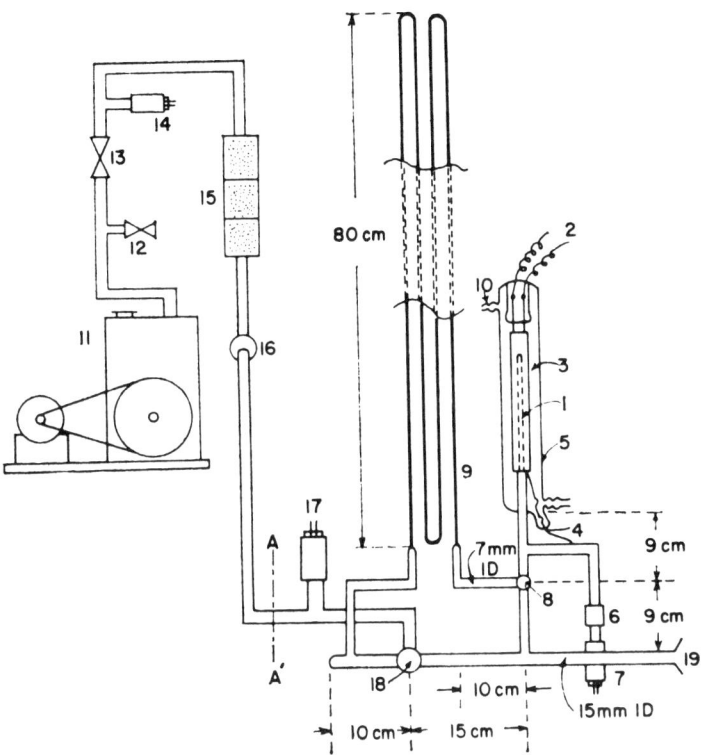

FIG. 1. The vacuum system (see Experiment 2.3).

nitrogen plus oxygen mixtures with varying concentrations over the silver. Because there is no indication that the nitrogen is participating in any way in the permeation process (2,3), N_2+O_2 mixtures offer a convenient tool for the study of permeation at oxygen pressures below 1 atm.

To measure the permeation at different oxygen pressures, the upper end (10) of the jacket is connected to a line which leads to a set of gas cylinders with different mixtures of N_2+O_2. It is recommended that the gas pass through a flowmeter and a microporefilter before it enters the jacket. A flow of about 450 cc/min has been found necessary in order to obtain satisfactory results. For measurements with air in the jacket, it is sufficient to have a membrane pump connected to inlet (10). The ends of the silver tube seem to be somewhat cooler than the middle and it is therefore recommended that the jacket be covered with aluminum foil. This gives better temperature uniformity along the heating element.

EXPERIMENTAL PROCEDURE

To start the experiment the same operating instructions are observed as described in Experiment 2.3. When taking measurements, it is practical to start with the highest temperature for the silver tube and work down to the lower ones. Table 1 gives a set of readings in Columns 1 and 3 as measured with the system described here, which can be evaluated for the activation energy of the permeation of silver by oxygen. For an evaluation, one first computes the throughput for different pressures. This is done with $P_1 = 0$. The result is listed in Column 4 of Table 1. For convenience, a graph has been plotted which gives the throughput as a function of the measured pressure (Fig. 2). The constants used for the computation of the throughput are given in this figure.

TABLE 1
PERMEATION OF OXYGEN THROUGH A SILVER TUBE

Temp. (°C)	1000/T (1/°K)	Pressure at capillary entrance (mTorr)	Flow of oxygen (mTorr·liter/sec)
Oxygen pressure in outer jacket: 745 Torr			
701	1.028	1050	0.67
677	1.052	900	0.53
664	1.068	735	0.38
622	1.118	440	0.18
Oxygen pressure in outer jacket: 156 Torr			
711	1.017	750	0.40
674	1.057	520	0.22
638	1.099	310	0.115

4.4. PASSAGE OF OXYGEN ACROSS A SILVER BARRIER

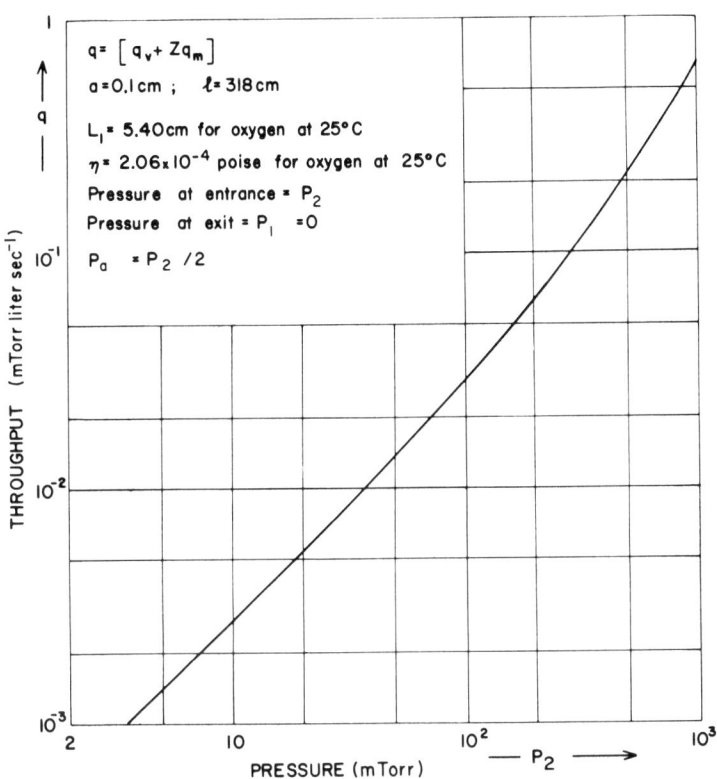

FIG. 2. Capillary throughput versus pressure at entrance.

When the logarithm of the flow across the silver barrier (measured as throughput) is plotted as a function of the reciprocal of the absolute temperature, with the pressure in the outer jacket as a parameter, one obtains a set of parallel straight lines. Figure 3 shows two of these lines for pressures of 745 and 156 Torr. The slope of these lines according to Eq. (2) is the activation energy of the permeation process and is found to be about 28,300 cal/g atom. The accepted value is 22,600 cal/g atom(4). The reason for this discrepancy will be discussed elsewhere.

From Fig. 3 one realizes easily that, with the temperature kept constant, Eq. (4) is fulfilled: $Q_1/Q_2 = (745)^{1/2}(156)^{1/2}$. This indicates that oxygen diffuses as an atom in silver.

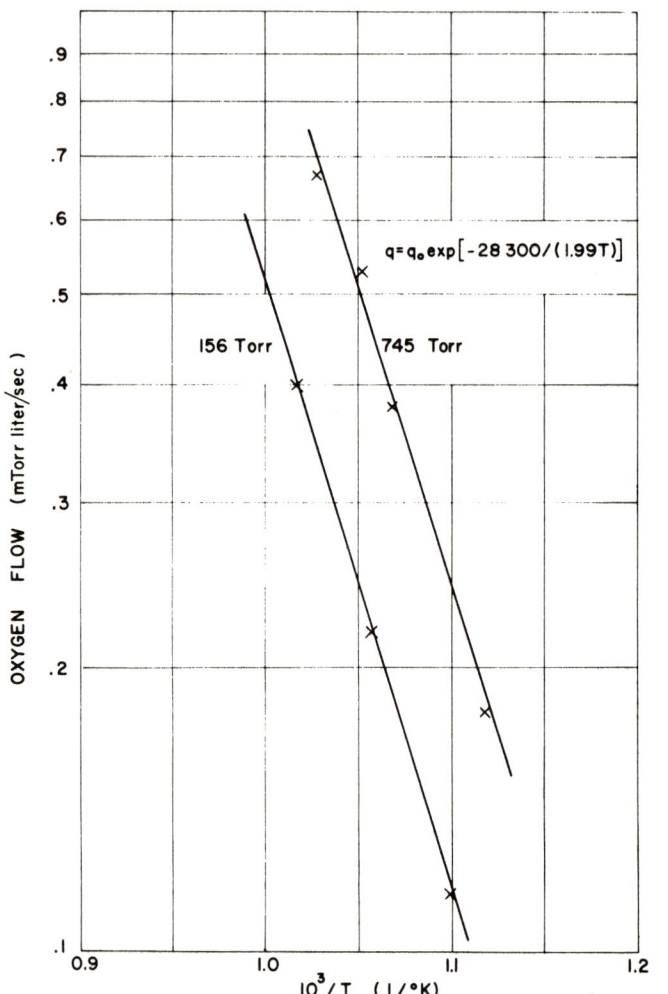

FIG. 3. Permeation of silver by oxygen.

If necessary, the Pirani gauge can be calibrated by a McLeod gauge connected to the ball joint (19). For permeation experiments, the female part of the ball joint is closed by a male counterpart. Calibration of the Pirani gauge for air is rather close to that for oxygen, as suggested by Ubisch(5). One can therefore use the gauge for measurements of oxygen pressures as well as for air pressures without applying conversion factors.

4.4. PASSAGE OF OXYGEN ACROSS A SILVER BARRIER

REFERENCES

1. K.M. Busen, Am. J. Phys., 35, Part 1, B398 (1967).

2. F.M.G. Johnson and P. Larose, J. Am. Chem. Soc., 46, 1377 (1924).

3. N.R. Whetten and J.R. Young, Rev. Sci. Instr., 30, 472 (1959).

4. C.J. Smithells and C.E. Ransley, Proc. Roy. Soc. (London), A150, 172 (1935), 182, Table V.

5. H.V. Ubisch, Anal. Chem., 24, 931 (1952).

Editorial Comment (VJH). A late contribution to the book was received from Chul Chu Lee, Yonsei University, Seoul, Korea entitled "The Measurement of Hydrogen Permeation Rate through a Palladium-Silver Tube for Ion Source." Since this contribution is very similar to Experiment 4.4., the reader is referred to the Journal of the Korean Physical Society, 1(1), 34 (1968) for details of the experiment.

Section 5

PROCESSES REQUIRING A VACUUM ENVIRONMENT

Experiment 5.1

THIN FILM EVAPORATION

M. T. Thomas

Department of Physics
Washington State University
Pullman, Washington

INTRODUCTION

Thin films of metals, semiconductors, and insulators have found many applications in industry and in particular the electronic industry. Also the basic physical and electrical properties and growth mechanism (epitaxy) of thin films are the subject of intensive fundamental research. For a review of some of the current thin film research areas, the interested reader should read some of the articles in the series, Physics of Thin Films(1), and look at the journal, Thin Solid Films(2). A good review of the use of thin films in the electronic industry is given in the book Thin Film Microelectronics(3). The purpose of this experiment is to become familiar with the basic techniques of metallic thin film vacuum evaporation and to investigate some of the parameters that influence the physical properties of the films.

EXPERIMENTAL SYSTEM AND PROCEDURE

The production of metallic thin films can be accomplished with a minimum of equipment. Also, the texture, adhesion, and electrical resistivity of thin films can be investigated with equipment that is readily available or easily obtainable. Some of the basic techniques and apparatus required for the production of thin films are briefly outlined below. For a more complete discussion of apparatus and techniques the reader is referred to the books by Strong(<u>4</u>) and Holland(<u>5</u>).

A. <u>EQUIPMENT</u>

1. Vacuum system: Any type capable of attaining a pressure of about 10^{-5} Torr

    (a) Two high current electrical feedthroughs for evaporation source heater (one plus base plate in bell jar system)

    *(b) Two electrical feedthroughs for substrate heater

    *(c) Two electrical feedthroughs for monitoring film resistivity

    *(d) One rotary feedthrough for shutter operation

2. Material for evaporation: metals such as Al, Sn, Ag, Au, Pb, Ge, etc. are easy to deposit.

3. Evaporation source heater filaments: tungsten wire helical-shaped baskets (or other shapes). These are commercially available or they can be easily fabricated by twisting two or three 15-mil tungsten wires together and then into the shape desired(<u>4</u>). Tantalum wire can also be used as a heater filament for most of the above metals.

4. Substrates: 1 in. x 3 in. or 1 in. x 1 in. microscope slides. Other material such as quartz, sapphire, or ceramics can also be used.

5. Heater filament power supply: high current autotransformer or a high current transformer controlled by a Variac.

6. Substrate holder: home built, the configuration of which depends on the geometry of the vacuum system. The baseplate in a bell jar system can be used to support the substrates with the evaporation source above the substrates.

---

*These items are optional and are not necessary for the initial experiments discussed below. However, if some of these items are available, additional experiments can be done.

5.1. THIN FILM EVAPORATION

*7. Substrate heater: home built

8. Detergent and organic solvents: for cleaning substrates. Solvents such as acetone and methanol are good.

9. Balance: for determining the amount of material deposited on the substrate. From this information a fairly good estimate of the thickness of the film can be obtained.

10. Scotch tape: for adhesion test

*11. Optical microscope: for examining film texture. This can be done to some degree without the aid of a microscope. A simple eye loupe would help. With a monochromatic light source interferometric thickness measurements can be made.

*12. Resistance measuring apparatus

13. Dark glasses (density of 4 or 5) or cobalt colored glass: for viewing glowing heater filament. This is important for eye safety.

B. FILM PRODUCTION

Any type of vacuum system that can attain a pressure of about 10^{-5} Torr can be used for these experiments. The system should be capable of being opened and closed easily since it will be instructive to make films under a variety of deposition parameters. Figure 1 shows a typical bell jar arrangement with the minimum equipment needed. Of course, a particular system arrangement will depend on the work chamber space and the materials available.

The process of producing a thin film is very simple and is briefly outlined below. A check of the geometrical arrangement should be made to ensure that electrical connections are tight and the substrate-to-heater-filament distance is properly set. The helical tungsten basket is filled with a charge of the material to be evaporated, the substrates cleaned and loaded in the holder, and the system closed and evacuated. Care should be taken to ensure that the charge of material is clean, i.e., degreased and dry, to minimize the substrate and film contamination. This is particularly important if there is no shutter in the system which would allow degassing and predeposition to vacuum clean the evaporant. Generally the system is allowed to reach its base pressure before a deposition is carried out. Once a suitable pressure is attained, the filament is slowly heated. If the system has a shutter it should be covering the substrates at this point. Continue to increase the filament temperature until the charge of material melts and wets the tungsten wire.

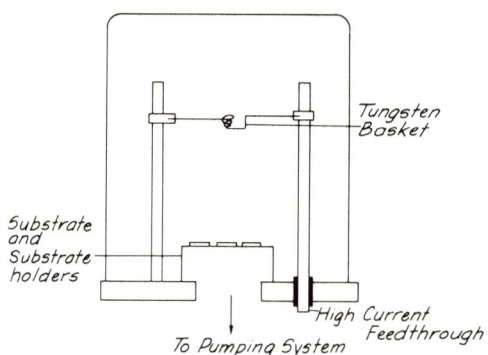

FIG. 1. Bell jar evaporation system.

Important: Be sure to use the dark, eye-protective glasses when viewing the glowing filament. With further increase in the temperature the material will begin to evaporate and a layer will begin to deposit on the substrates. For those systems with a shutter, the shutter should be opened after a few seconds of deposition to expose the substrates. After the film deposition is completed, allow the filament and film to reach room temperature before opening the vacuum chamber.

OBSERVATIONS

Thin film texture, film adhesion to the substrate, and the film resistivity are properties of deposited films that can be investigated as a function of deposition parameters with a minimum of equipment. A visual inspection of the appearance of the film is all that is necessary to characterize the texture. Determine if the film has a shiny or cloudy appearance, if it is smooth, wavy, or granular (grainy), and if it has a uniform texture. A low power optical microscope is helpful especially in determining granularity. To test the adhesion of the film to the substrate, the scotch tape test is used. Apply the tape to the film and see how readily the film peels off the substrate when the tape is stripped off. The resistivity of the film can be calculated by the total resistance of the film and its dimensions. The length and width are determined by the substrate, but the thickness must be measured. One of the simplest methods of estimating the thickness of an evaporated film is to weigh the substrate before and after deposition. The thickness of the film is then determined from the equation $t = m/A\rho$ where t is the thickness in cm, m is the increase

5.1. THIN FILM EVAPORATION

in mass of the substrate in grams, A is the area of the substrate in cm^2, and ρ is the bulk density of the material being deposited in g/cm^3. Of course, if the equipment is available, interferometric methods can be used for thickness determinations (5).

Many deposition parameters change the properties discussed in the previous paragraph. The first and easiest ones to investigate are substrate preparation, rate of evaporation, and background pressure. If various kinds of substrate material such as quartz, sapphire, ceramics, NaCl, etc. are available, or if the substrate temperature can be varied, it would be instructional to investigate the effect of these parameters on the films. Only a few of the possible experiments and variations of combinations of the deposition parameters are mentioned, so it is hoped that other experiments will be tried.

ADDITIONAL EXPERIMENTS

The substrate preparation is important in controlling the properties of the film. Pick a given set of deposition parameters: Source-to-substrate distance, evaporation rate (controlled by filament temperature), background pressure, and substrate temperature. Mark one side of the glass slides and then clean them in various ways. For example, one could have no cleaning at all, another could be brushed clean using the fingers, use detergents and organic solvents on others. Scratch a slide and place a thumb print in the center of another. What do the films deposited on these substrate look like? What is the effect of these various treatments on the texture, adhesion, and resistivity? What happens to the film if there are dust particles on the substrate? Is the substrate preparation as important for thick films as for very thin films?

Use the substrate cleaning procedure which produces shiny and smooth films. Investigate the effect of film thickness on adhesion. Also it would be interesting to make films at different pressures, for example from 5×10^{-3} Torr to 1×10^{-5} Torr or the lowest attainable pressure of the system. How do the texture, adhesion, and resistivity vary as a function of pressure? If the resistivity changes, why does it change? At a high pressure does the deposition rate change the resistivity?

There are many permutations of film and deposition parameters which can be varied. The important thing is to think about what some of the important parameters are and study them. If apparatus is available to examine film structure or composition, the number of experiments is almost unlimited.

REFERENCES

1. Physics of Thin Films (G. Hass and R.E. Thun, eds.) Vols. 1 - 5, Academic, New York, 1963-1968.

2. Thin Solid Films (L. Holland and J.A. Dillon, Jr., eds.) Vol. 1 to current volume, Elsevier, Amsterdam, Netherlands, 1967 to present.

3. Thin Film Microelectronics (L. Holland, ed.) Wiley, New York, 1965.

4. J. Strong, Procedures in Experimental Physics, Prentice-Hall, Englewood Cliffs, N.J., 1938, Chap. 4, p. 151.

5. L. Holland, Vacuum Deposition of Thin Films, Wiley, New York, 1956.

Experiment 5.2

FABRICATION OF A NICHROME RESISTOR

R. P. Riegert and G. Breitweiser

Sloan Instruments Corporation
Santa Barbara, California

INTRODUCTION

Deposited thin film resistors play an important role in the microelectronics industry. Resistors are deposited on both ceramic substrates (as for thin film passive networks) and on silicon integrated circuits (actually on a silicon dioxide layer). Nichrome resistors have met with wide acceptance, partly because of the tenacity with which the chromium adheres to the substrate. Sheet resistance values in the range from 40 to 800 Ω per square can be attained, the most popular being about 250 Å thickness, giving about 200 Ω per square.

This experiment is designed to illustrate a practical nichrome resistor process, the use of resistance sources, masking techniques, quartz crystal monitoring, conductance monitoring, and use of a multiple beam interferometer. The departure of the resistance versus thickness relation from linearity in the very thin film region serves to illustrate influence of nucleation, island structures, and the growth process in general.

EXPERIMENTAL SYSTEM

A conventional evaporation system can be used for this experiment. This consists of a bell jar and a base plate mounted above the main high vacuum valve of a liquid-nitrogen trapped-oil diffusion pumping system. The base plate should have some high current feedthroughs for evaporation of aluminum and nichrome and several other instrumentation feedthroughs.

EXPERIMENTAL PROCEDURE

1. Clean several 1 in. x 3 in. x 1 mm glass microscope slides using a detergent solution in distilled water. Rinse in distilled water and dry under infrared lamps.

2. Deposit 1500 Å aluminum contact electrodes using a stranded tungsten wire spiral basket source. Clip two slides together (using an alligator clip) to mask a 1-in. strip across the center of the slide (see Fig. 1). Use a quartz crystal monitor to control deposited aluminum to 1500 Å. Keep the slides and monitor about 14 to 18 in. above the source.

3. Prepare a resistor mask using a 35 x 50 mm Corning microscope cover glass. Scribe the cover glass lengthwise down the center and break along the scribe line. Clip the two halves to the substrate slide to leave a gap about 1/8 in. wide running lengthwise between the pads. (See Fig. 2.)

4. Clip an electrical lead to each of the aluminum pads. Feed these leads outside the chamber and connect to monitor conductance of the deposition resistor. Use a 1 V dc supply and monitor current using an electrometer with several ranges. Connect a strip chart recorder to record the electrometer reading (current).

5. If a dual pen recorder is available, connect the second pen to record frequency change of the crystal monitor. Otherwise, note frequency data on the conductance strip chart during the deposition.

6. Using a heavy stranded tungsten filament, evaporate nichrome (80 Ni, 20 Cr) onto the slide through the mask. Record conductance and frequency change data. Maintain a rate of 15 to 50 Å/min to a total thickness of 350 Å. Change current scales on the electrometer as the deposition proceeds.

7. Accurately measure the length and width of the resistor strip (between the aluminum electrode areas).

8. Overcoat the resistor slide with approximately 1500 Å of aluminum and measure the nichrome thickness with an interferometer.

5.2. FABRICATION OF A NICHROME RESISTOR

9. Using this thickness value and the total frequency change in the nichrome deposition, convert the frequency data into thickness.

10. Plot conductance and resistance versus thickness.

11. Calculate the nichrome sheet resistance in ohms per square and resistivity in ohm cm.

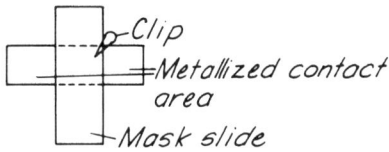

FIG. 1. Deposition of contacts.

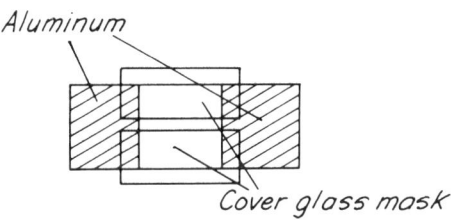

FIG. 2. Preparation of resistor.

BIBLIOGRAPHY

1. W. Liben, et al, *Microelectronic Engineering: Fabrication Technology*, John Hopkins University, (November 1965), AD 624-315-: available from Clearinghouse for Federal Scientific and Technical Information, U.S. Department of Commerce.

2. L. Holland, *Vacuum Deposition of Thin Films*, Chapman and Hall, London, 1966.

3. C.D. Simmons, *SCP & Solid State Technology*, 21, (March 1964).

4. L.I. Maissel, *Solid State Technology*, 27, (May 1968).

Experiment 5.3

SPUTTERING

P. Grosewald

IBM Watson Research Center
Yorktown Heights, New York

INTRODUCTION

Prior to 1964, sputtering was a laboratory technique studied only as a means of obtaining information about basic physical mechanisms, although the phenomenon of sputtering was first observed over a century ago(1).

Renewed interest in sputtering as a process arises from the exploding technologies of film deposition; for coatings of such varied objects as jet turbine blades and razor blades and for the deposition of resistive, conductive, semiconductive, and insulating thin films in the expanding microelectronics industry. Comparisons with other techniques must be left to other specialized works (2-6), but sputtering does have certain advantages for depositing films, e.g., low temperature deposition of refractory materials such as tungsten and improved film-substrate adhesion.

Sputtered films are similar to evaporated films; in fact, sputtering has been called impact evaporation, since the basic mechanism involves ion bombardment and ejection or sputtering of material to be deposited.

Positive ions from a gaseous plasma (cloud of gas atoms, ions, and electrons with a net charge of zero) are accelerated through an electron-free region (Langmuir Sheath) to bombard a cathode of target material. These highly energetic ions have sufficient energy to knock off or "sputter" atoms from the surface of the target. The precise mechanism is unclear, but it is a physical process rather than a chemical one. The momentum exchange theory explains that binding energies of surface atoms are overcome and these are atoms knocked out of the target via either an ion rebounding from within the target or by a chain reaction transmitted by the lattice.

The materials thus removed impinge on all nearby surfaces. If conditions are suitable, condensation takes place.

In diode sputtering, the relatively high potential between the cathode and ground plate, at pressures between 10^{-1} and 10^{-2} Torr, causes gaseous breakdown and generation of a "glow discharge" (see Fig. 1).

In triode sputtering, a hot filament and an anode are added to the diode system, thus separating the diode plasma formation and sputtering functions. This arrangement permits ionization at lower pressure.

Reactive sputtering involves a combination of a sputtering material with a constituent of the gaseous plasma, either at the cathode, in the gaseous phase, or via a reaction at the substrate (the latter being most probable).

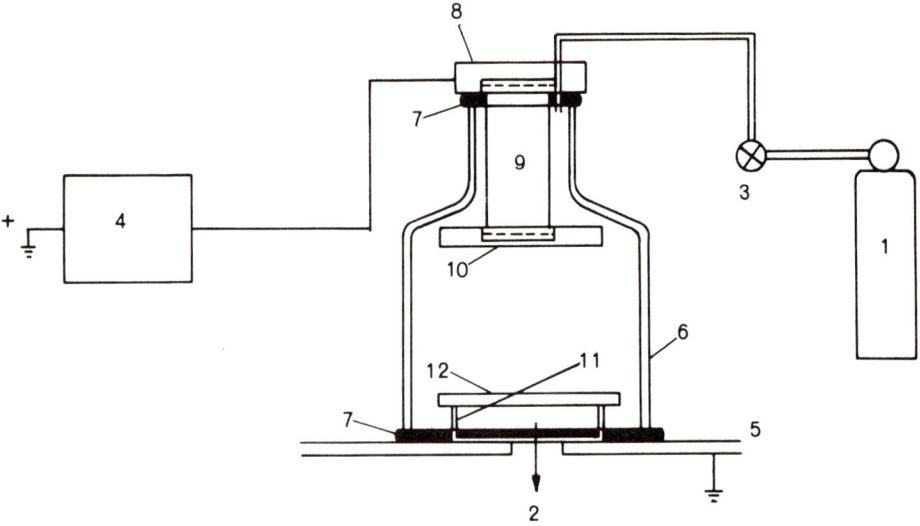

FIG. 1. Diode sputtering system. 1. argon cylinder, 2. vacuum system, 3. needle valve, 4. H.V. supply, 5. metal base plate, 6. chamber walls, 7. seal, 8. metal plate, 9. holder, 10. target, 11. support, 12. substrate.

5.3. SPUTTERING

In the case of multicomponent sputtering, there is some evidence(7-10) that sputtering yields molecular "chips" as well as atoms. There is also very strong evidence that after a brief conditioning period, multicomponent stoichiometry is maintained during sputtering(3,8,11).

Sputtering involves a number of variables. Control of the sputtering process is achieved by adjusting these parameters to yield deposits (Table 1).

Gas pressure directly controls the plasma density and thus the ion current, but also retards the mean free path of the sputtered material. Lower pressures imply that more atoms hit the substrate per unit time with higher (average) energy resulting in greater uniformity and in reduced strain in the films.

The cathode (target)-substrate distance is important in that too great a distance causes the velocities of the sputtered material to randomize and decrease due to collisions; while too small a distance can often emphasize the directionality of sputtering, leading to nonuniform (thickness) or annular deposited films.

For diode or two electrode sputtering, the thickness (t) of the deposited film can be approximately expressed as:

$$t = \frac{K I_c \, time}{p d_{cs}}$$

where d_{cs} is the cathode (target)-substrate distance, p is the gas pressure, I_c is the cathode ion current, and K is a constant, dependent on materials and geometry.

I_c is governed by p and the cathode potential, V_c; K is determined by the materials used and the system geometry. The thickness of the film is governed by exposure time and sputtering potential; substrate temperature and reactive gas additions to the system determine the composition and structure of the film.

EXPERIMENTAL SYSTEM

An unsophisticated dc diode sputtering system (Fig. 1) can be built from the equipment listed in Table 2. Items 5, 8, and 9, are best made of conductive materials which do not sputter at an appreciable rate. Wehner and his co-workers(12) have published most of the data on sputtering rates (as a function of controlling variables) for a wide variety of materials. Probably the best material is either

TABLE 1

Sputtering Parameters

---

a) Gas pressure

b) Gas species (ionic)

c) Plasma density

    1) Gas pressure

    2) Filament current

    3) Anode current

    4) Magnetic field

d) Cathode (target) - substrate separation

e) Cathode (ion) current

    1) Plasma density

    2) Target conductivity (h)

    3) Target voltage (f)

f) Cathode (target) potential

    1) Plasma density

    2) Target conductivity

g) Target temperature

h) Target material

i) Target geometry

---

aluminum or stainless steel. As an alternative, these items may be formed of copper; insulation must then be used to avoid formation of a discharge (and thus sputtering) outside the region between the target electrode and the substrates.

The target material (Item 10) may be fabricated from a solid piece, a sintered disk or a foil wrapped around an aluminum or stainless disk. A 1 to 3 kV potential yields a satisfactory glow discharge across a 1 to 3 in. target-ground gap.

5.3. SPUTTERING

TABLE 2

Equipment

---

1. Argon cylinder
2. Vacuum system
3. Gas flow metering - needle valve
4. 3 kV, 200 mA power supply
5. Metal plate
6. Glass reducer or glass pipe (O-ring grooves in either end)
7. Two O-rings for #6
8. Metal disk, tapped for 9
9. Metal (Al) rod, threaded either end, to be fastened to 8, and to hold target
10. Target disk, drilled and tapped to mate with 9, the cathode
11. Glass cylinder (substrate holder)
12. Microscope slide (substrate)

EXPERIMENTAL PROCEDURE

1. The chamber is evacuated to a pressure of less than 10^{-3} Torr and preferably to less than 10^{-5} Torr.

2. Argon and/or other gases are then admitted to the chamber via the needle valve.

3. The flow rate is balanced against the pumping speed until equilibrium is reached at 5 to 10 µ.

4. A high potential (~1 kV) is applied between the target and baseplate.

5. The chamber pressure is raised, by throttling the high vacuum valve, until a glow discharge occurs (20 to 70 µ).

OBSERVATIONS

The following series of experiments can be used to verify the formula given above.

1. The needle valve setting can be varied, thus changing the pressure. The operator should note the change in size of the nonglow region near the cathode, the cathode dark space. Holding the potential, the deposition time, and separation constant, the thickness and the target current (I_c) will be found to have changed. Similarly, changing the potential or the target-substrate distance with the other parameters constant will cause the final thickness to vary. It is suggested that copper be used as a target in this series of experiments as well as the next one.

2. The criticality of the spacing between the target and substrate may be illustrated by varying this distance. It will be seen that at too great a spacing, film thickness will be greatest below the target center and will decrease radially. As the separation decreases, the film thickness will become more uniform. As the distance decreases still further, the dark space will grow and an annular deposit will result, with minimum center thickness.

3. The last series of experiments are best done with materials such as tantalum or tungsten. Films of these materials are not produced readily by any other method. As a first step, films are deposited on a substrate and the thickness and resistivity are determined. Keeping all parameters the same, other inert gases can be substituted for the argon. It will be seen that incident ion energy has an effect on the thickness. Finally, oxygen may be added to the sputtering gas species. It will be found that oxygen decreases the sputtering rate and increases the sheet resistivity. It is suggested that the oxygen content be changed from run to run. 1%, 5%, 10%, 20%, and 30% of the total gas content will show the dependency of thickness and resistivity on reactive gas species.

4. The effects of pressure, target potential, cathode-substrate separation, target materials, and time can all be evaluated with this system. The use of different substrate materials permits

5.3. SPUTTERING

investigation of additional parameters affecting the formation of thin films. The use of masks (e.g., a penny) on the substrates and of photolithographic methods would also serve as an introduction to an expanding area of film technology.

REFERENCES

The following list of references is not extensive but Ref. 2 and, especially, Ref. 3 offer complete and up-to-date tabulation of the published literature.

1. W.R. Grove, Phil. Trans., B142, 87 (1952).

2. L. Holland, Vacuum Deposition of Thin Films, Wiley, New York, 1956.

3. L.I. Maissel, Deposition of Thin Films by Cathode Sputtering in Physics of Thin Films, Vol. 3, (G. Hass and R.E. Thun, eds.), Academic, New York, 1966.

4. P.D. Davidse and L.I. Maissel, J. Appl. Phys., 37, 574 (1966).

5. L.I. Maissel, 1st Conf. Elements of Sputtering, 1968.

6. E. Kay, 1st Conf. Advances in Electronics and Electron Physics, 17, (L. Marten, ed.), Academic, New York, 1962.

7. S.P. Wolsky, Trans. 10th Nat. Vac. Symp., 309 (1962).

8. G.J. Ogilvie, Aust. J. Appl. Phys., 31, 402 (1961).

9. W.R. Sinclair and F.J. Peters, J. Am. Cer. Soc., 46, 20 (1963).

10. A.R. Janus and G.A. Shirn, 1st Symp. on Deposition of Thin Films by Sputtering, Consolidated Vacuum Corporation, 1966.

11. N.C. Miller and G.A. Shirn, 2nd Symp. on Deposition of Thin Films by Sputtering, Consolidated Vacuum Corporation, 1967.

12. N. Laegreid and G.K. Wehner, J. Appl. Phys., 32, 365 (1961); G.K. Wehner and D. Rosenberg, ibid. 887 (1961); ibid, 33, 1842 (1962); G.K. Wehner, Phys. Rev. 108, 35 (1957); ibid, 112, 1120 (1958).

Experiment 5.4

EJECTION PATTERNS IN
SINGLE CRYSTAL SPUTTERING

G. K. Wehner

Department of Electrical Engineering
University of Minnesota
Minneapolis, Minnesota

INTRODUCTION

One phenomenon which is being actively studied at the present time and which has found a number of practical applications is the interaction of ions with solid surfaces. Even though a great deal of work has been done in this field, covering a period of many years, basic understanding of the mechanism involved is only now beginning to emerge. Ion bombardment cleaning of solid surfaces, either in a general glow discharge of a noble gas or with a beam of ions has been used as a first step to prepare materials for industrial processing or for fundamental scientific studies. Sputtering of thin films for the fabrication of thin film devices is being used extensively by the thin film industry. Ion beams are also used as a probe to study solid surfaces and also to investigate some properties of bulk solids. A good review of this field, which includes an extensive bibliography, is found in the book <u>Atomic and Ionic Impact Phenomena on Metal Surfaces</u> by Kaminsky.

The following experiment is a fairly simple introduction to the field of ion-solid surface interactions. There are a number of variations and extensions to this experiment which can be easily pursued if one wishes to learn more about the experimental techniques and problems associated with this type of research.

EXPERIMENTAL SYSTEM AND PROCEDURE

Establish in a demountable glass bell jar vacuum system, a low pressure argon plasma between a thermionic cathode and an anode. Typical data: argon gas pressure, 1 to 5 mTorr; discharge current, 1 A; anode-cathode voltage drop, 40 V. Cathode can be taken from a commercial thyratron.

Insert in this plasma a metal single crystal sphere (e.g., silver). Apply to the sphere from a separate power supply a negative voltage of the order 100 to 1000 V with respect to anode. Bombarding ion current density should be of the order of 1 mA/cm^2 or higher. For achieving this it might be necessary to concentrate the plasma in the vicinity of the sphere by means of an internal permanent magnet. Provide a glass plate or cylinder or half sphere in target vicinity for catching the sputtered metal atoms and observe the deposit patterns which will appear after sputtering times of the order of hours.

BIBLIOGRAPHY

1. M. Kaminsky, Atomic and Ionic Impact Phenomena on Metal Surfaces, Academic, New York, 1965.

2. G.K. Wehner, Phys. Rev., 102, 690 (1956).

3. G. Carter and J.S. Collingon, Ion Bombardment of Solids, Heinemann Educational Books, Ltd., 1968, Chap. 5.

Section 6

SPECIAL PROJECTS

Experiment 6.1

STUDY OF THE SUBLIMATION
OF ICE AT VARIOUS PRESSURES

William N. Parker

RCA Electronic Components
Lancaster, Pennsylvania

INTRODUCTION

Vacuum systems are frequently used for the freeze-drying (sublimation) of frozen foods and pharmaceuticals(1,2). Moisture is removed moved by sublimation if the partial pressure of water vapor over the frozen material exceeds that in the surrounding vacuum. If this partial pressure is greater than about 4 Torr, however, the ice melts and true freeze-drying does not take place. Typically, the necessary low partial pressure is maintained by keeping the total system pressure well below 1 Torr. If microwave dielectric heating is used to supply the heat of sublimation, the total maximum allowable pressure may be as low as 50 mTorr to prevent glow discharge(3,4). Adequate pumping systems for such pressures are expensive.

Sublimation at higher (even atmospheric) total pressures is theoretically possible as long as the residual gas is sufficiently dry(5). Such systems should be much cheaper to operate(6). Even the glow discharge problem disappears at pressures above about 30 Torr(7). However, practical sublimation at high pressures can involve several unexpected effects. The experiment described here vividly demonstrates such effects as self-convection, laminar flow, radiant heat transfer, importance of mean-free-path in trap design, insulating ice on the condenser, thermal shock, and whisker growth.

EXPERIMENTAL SYSTEM

A glass cylinder (or "spool") about 7 in. in diameter and 9 in. long rests on a flat plate and supports a liquid-nitrogen chevron trap acting as a moisture condenser; Fig. 1. The plate contains an orifice which is connected to a mechanical vacuum pump through throttle valve 1 for evacuation of the glass cylinder. Pressure from 100 mTorr to more than 100 Torr may be measured by use of both thermocouple and Bourdon gauges. Throttle valve 2 may be used as a controlled leak and to break the vacuum. Inside the glass cylinder the preweighed ice-cake sample rests on a 300 W strip heater which in turn is supported by a urethane insulation block. A measured amount of heater power (up to about 15 W) is supplied to the heater from a Variac. Power for the heater passes through two thin copper strips sandwiched between a pair of well greased rubber vacuum gaskets at the lower end of the glass cylinder. A single gasket is sufficient at the upper end.

EXPERIMENTAL PROCEDURE

Sublimation tests are run at atmospheric pressure, 76 Torr, and 200 mTorr.

For each test, a frozen ice-cake sample (about 100 g) is weighed and placed on the cool heater inside the glass cylinder. The trap is kept cold with liquid nitrogen before and during the tests. The ice sample tends to cool off as a result of loss of heat of sublimation and there is a consequent reduction of vapor pressure. For effective sublimation, the sample temperature should be only slightly below freezing. The necessary added heat can be supplied by the heater. Excessive heat causes the ice-heater interface to start melting; this effect can be observed if the ice cake is clear. If heat is supplied too suddenly, thermal shock may crack the ice.

A careful record of power input as a function of time is helpful in determining the total W/min energy input for comparison with the theoretical energy represented by the weight lost by sublimation at the end of each test (49 x g lost = equivalent W x min).

The test at atmospheric pressure produces a vigorous convection of fog in a downward direction. Thick frost usually collects on the outside of the glass cylinder as a result of convective cooling. The weight by sublimation is much less than the equivalent W/min input because of heat loss by convection and radiation from the ice-cake sample.

6.1. SUBLIMATION OF ICE AT VARIOUS PRESSURES

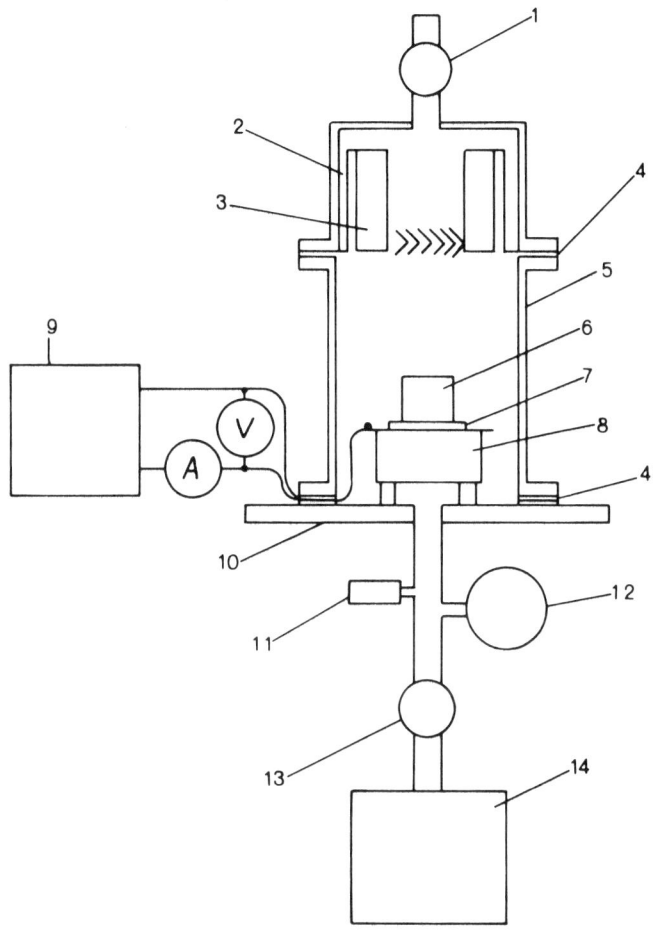

FIG. 1. Apparatus for sublimation tests. 1. valve, 2. chevron trap, 3. liquid nitrogen reservoir, 4. gasket, 5. glass cylinder, 6. ice cake sample, 7. strip heater, 8. insulation block, 9. Variac, 10. base plate, 11. T.C. gauge, 12. Bourdon gauge, 13. valve 1, 14. mechanical pump.

Operation at 76 Torr (about 27-in. vacuum) may be obtained by partially pumping down and then closing valve 1. Convection-cooling effects are much less and the glass-cylinder outside wall remains clear and dry. However, visible clouds of vapor (fog) can be observed coming down from the condenser. Moisture accumulates on the condenser as fluffy "snow." Some of the long strings tend to drop off and deposit flakes back on the ice cake which seem to require a long time to re-evaporate. Very little heat can be applied without melting at the interface, and any water droplets so formed persist quite a while. (Refreezing may be accomplished by temporarily pumping down to less than 1 Torr and then bleeding back to 76 Torr through valve 2.) The slow evaporation of water probably indicates that the air coming down from the condenser is quite warm and not very dry because of the insulating effect of the fluffy snow. The weight loss resulting from sublimation often exceeds the equivalent electrical energy input; this effect indicates considerable heat pickup by the sample by radiation from the glass walls.

Even when the 76 Torr operating pressure is achieved by continuous bleeding and pumping, visible moving clouds of fog can be observed. Apparently the incoming room air does not become dry and cold in passing through the chevron baffle of the liquid-nitrogen cooled trap. The very short mean-free-path allows essentially laminar flow through the baffle (unless very close spacing is used).

Sublimation at 200 mTorr is quite conventional and free of the effects observed at higher pressures. At the start of sublimation, peculiar "whiskers" grow out from the ice-cake sample. These growths may later disappear. The condensed ice layer is also quite dense. The sample weight loss usually exceeds the equivalent electrical energy input as a result of heat radiation to the sample from the glass walls.

BIBLIOGRAPHY

1. J.C. Harper and A.L. Tappel, Advances in Food Research, 7, 171, Academic, New York, 1957.

2. R.F. Burke, and R.V. Decareau, Advances in Food Research, 13, 1, Academic, New York, 1964.

3. S. Jackson, S. L. Richter, and C.O. Chichester, Food Technology, 11(9), 468, (1957).

4. M.W. Hoover, A. Markantonatos, and W.N. Parker, Food Technology, 20(6), 103, (1966).

5. H.T. Meryman, Freeze-Drying of Foods (R.F. Fisher, ed.) National Research Council, Washington, D.C., 1962.

6.1. SUBLIMATION OF ICE AT VARIOUS PRESSURES

6. T.H. Woodward, Quartermaster Contract Rept. (DA-19-129-QM-1597), 1961.

7. W.N. Parker, Microwave Power Engineering, $\underline{2}$, 238, Academic, New York, 1968.

Experiment 6.2

STUDY OF FRICTION

P. E. McElligott

General Electric Company
Schenectady, New York

INTRODUCTION

The development of improved vacuum techniques is, in part, responsible for recent progress in our understanding of the friction, lubrication, and wear of materials. This is because the behavior of materials in sliding contact is usually sensitive to the chemical nature of the sliding surfaces. Vacuum techniques play a useful and often critical role in controlling the chemistry of sliding surfaces, which under experimental conditions may range from atomically clean to deliberately contaminated in a controlled manner.

The sliding of one solid over another is resisted by a frictional force which is dependent on the nature of the sliding materials. For convenience, the coefficient of friction:

$$\mu = \frac{\text{Frictional force}}{\text{Load}} \tag{1}$$

is often used to describe the effect of the frictional force. A distinction is usually made between static friction (μ_s), or the force required to initiate observable sliding motion, and kinetic friction (μ_k), which is the force required to maintain sliding. For most systems $\mu_s > \mu_k$.

Early observations on the frictional behavior of materials were made by daVinci, Amontons, and Coulomb, among others. These observations are summarized in the so-called Classical Laws of Friction:

1. The frictional force between two sliding bodies is independent of the apparent area of contact between the bodies.

2. The frictional force is directly proportional to the load on the bodies (μ is constant).

3. The value of the kinetic friction is approximately independent of the sliding velocity.

For many years friction was thought to result from the mechanical interlocking of the surface asperities which are always present regardless of the degree of polish of the sliding surfaces. An alternative theory to this roughness hypothesis, namely, that friction is the result of adhesive forces at the sliding interface, was rejected on the grounds that it could not explain the independence of the frictional force and the apparent area of contact. However, the adhesion hypothesis has increasingly been shown to be the more correct of the two theories. As a result, the classical laws of friction may be regarded as somewhat analogous to the laws of classical mechanics; that is, useful in describing the more macroscopic behavior, but yielding place to the adhesion theory (or quantum mechanics) when the interactions between bodies are examined in sufficient detail.

Bowden and Tabor[1], in their interpretation of the adhesion theory, assume that the friction of metals results from two separate processes during sliding contact. The first process involves the formation, growth, and eventual shearing of metallic junctions formed across the interface. The second process, which can occur when one material is measurably harder than the other, involves the ploughing of the softer material by asperities on the surface of the harder materials. In many cases the shearing term is the larger, and when dealing with the friction between identical metal surfaces the ploughing term may be neglected due to the equal hardness of the materials.

The asperities present on all metal surfaces play a critical role in the adhesion theory of friction. When two metal surfaces are brought together, the initial contact is made on mating asperities, and the real area of contact, A_r, is much smaller than the apparent or geometric area. Elastic deformation occurs in the mating asperities, followed by plastic deformation of the most highly stressed asperities as the load is increased. The plastic deformation enables an additional set of somewhat lower asperities to make contact and thus help support the load.

For a given load, L, the real area of contact between two materials may be approximated by:

$$A_r = \frac{L}{p} \tag{2}$$

where p is the yield pressure or penetration hardness of the softer

6.2. STUDY OF FRICTION

material. The shearing term of the friction force may then be written as:

$$F = A_r \, \sigma = \frac{L\sigma}{p} \qquad (3)$$

where σ is the mean tangential stress required to shear the junctions formed by the mating asperities.

From Eq. (3) it is seen that the frictional force is not related to the geometric area of contact, but is directly proportional to the normal force or load. The dependence of F on the material properties occurs through the terms σ and p. Furthermore, because σ and p in many cases have a similar dependence on temperature, the ratio (σ/p) may be reasonably constant at both low sliding speeds (low surface temperatures). Thus, the adhesion theory of friction is able to explain many of the classical observations on sliding system.

Experiment A: <u>Investigation of the effect of oxide films on metallic friction</u>

Metal surfaces are normally covered with an oxide film and it is reasonable to expect the presence of the oxide to influence the friction through the terms σ and p. Experimentally it is observed that the hardness of the oxide relative to the hardness of the underlying metal is an important parameter in determining F. Oxide films in many cases can prevent metal-to-metal contact at mating asperities (at relatively light loads) and thus inhibit the cold welding and growth of the junctions. It is for this reason that oxidized metal surfaces frequently exhibit lower friction than the same metals after vacuum outgassing.

1. Determine μ for molybdenum surfaces in the "as received" condition (after washing and degreasing).

2. Determine μ after outgassing both molybdenum surfaces at 950°C in vacuum.

3. Determine μ after exposing both molybdenum surfaces to $\sim$1 mm O_2 at 500°C for 15 min.

4. Optional: Determine μ as a function of load and sliding speed.

Experiment B: <u>Investigation of the friction between two sliding molybdenum surfaces as it is affected by a solid lubricant generated in situ</u>

Under conditions of high loading and low speeds, conventional fluid lubricants such as mineral or vegetable oils frequently become ineffective. This is the so-called boundary friction regime where the sliding surfaces are separated by only a few angstroms, and where the low viscosity of a fluid loses its advantage. Solid lubricants in a variety of forms have been used successfully for many years.

The solid lubricant may be added mechanically (e.g., powdered graphite), or be an integral part of the sliding system in the form of an evaporated film of soft metal, or it may be generated in situ through a specific chemical reaction. Lubricants which contain specific additives to form a lubricating surface compound in situ are known as extreme pressure (E.P.) lubricants. Commercial E.P. lubricants usually contain either chlorine, sulfur, or phosphorus compounds as the active agents. In most cases the E.P. additives are thought to function by forming a low-shear-strength compound on the metal surface. Many of these surface compounds have a lamellar or layer structure in which it is relatively easy to cause shearing of adjacent, weakly bound crystal planes. Molybdenum disulfide (MoS_2) and ferric chloride ($FeCl_3$) are two such layer compounds which have good lubricating qualities, and which may be generated in situ on sliding surfaces. In addition to reducing the shear strength, the E.P. agents probably also reduce the amount of intermetallic contact by the surface asperities.

1. Determine μ after outgassing the molybdenum surfaces.

2. Determine μ after exposing the molybdenum surfaces to H_2S at room temperature.

3. Determine μ after exposing the molybdenum surfaces to H_2S at 500°C.

Comments. A vacuum friction experiment will generally require the introduction of motion and electric current into the vacuum chamber, and the manipulation of various gases. Because laboratory facilities and time schedules will vary, two techniques are described for the experiments. The techniques differ in degree of sophistication and equipment requirements, and their selection may be left to the instructor. Figures 1 and 2 illustrate the principles and apparatus involved, and give suggested dimensions.

The coefficient of friction for molybdenum sliding on molybdenum is expected to vary as shown below:

Molybdenum condition	Surface film	Approximate μ
As received, washed and degreased	oxide (~20 Å)	1
After vacuum outgassing for 15 min at 950°C		2-3

6.2. STUDY OF FRICTION

Molybdenum condition	Surface film	Approximate μ
After exposure to 1 Torr of oxygen for 15 min at 500°C	oxide (~120 Å)	<1
After outgassing and exposure to 1 Torr of H_2S at 300°C	MoS_2	0.2

Thus, for a given load, the required detection sensitivity for the friction force may be determined in advance.

EXPERIMENTAL PROCEDURES

CRITICAL-ANGLE METHOD

Figure 1 illustrates the critical-angle method of determining the coefficient of friction. Although this technique has the advantage of simplicity, it is particularly susceptible to vibration effects, and is only capable of measuring the static friction.

The critical angle is the inclination from the horizontal at which the rider just begins to slide. By resolving the forces, the load is shown to be $L = W \cos\theta$ and the friction force is $F = W \sin\theta$. Thus the coefficient of friction is obtained directly as $\mu = F/L = \tan\theta$.

In Fig. 1, a lever arm attached to a rotary motion feedthrough rotates the taut molybdenum wire about a horizontal axis. The angle of inclination may be measured directly by means of a pointer attached to the shaft or to one end of the lever arm. Alternatively, for higher sensitivity, a small mirror may be attached to the shaft and the angular deflection of a reflected light beam may be measured.

The molybdenum wire, insulated from the lever arm, is heated resistively by means of slack leads connected to an electrical feedthrough. It is desirable to separate the sliding surfaces during heating. Rotation of the lever arm to a near vertical position will enable the cylindrical rider to rest against the fixed platform in Fig. 1. The rider is then heated by radiation from the wire.

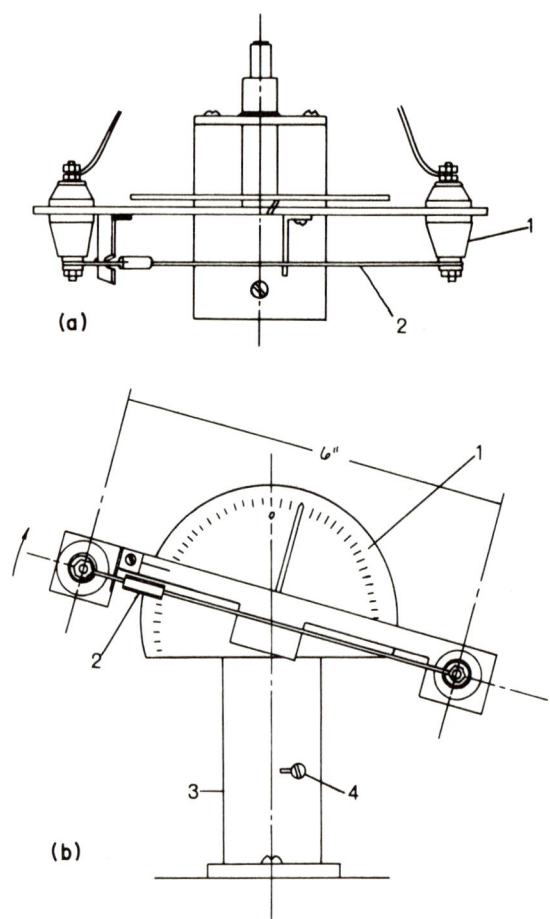

FIG. 1. Apparatus for critical-angle method. (a) Top view. 1. Brinbach-458 insulators, 2. 1/32" Diam moly wire. (b) Side view. 1. Ackerman dial, 2. 1/8" O.D. x 0.02" wall moly tube, 3. bearing tube, 4. centering finger for degassing. (c) This view shows apparatus position during degassing. 1. 1/4"-Diam stainless steel shaft-free fit couple to rotary motion vacuum feed, 2. all parts stainless steel.

6.2. STUDY OF FRICTION

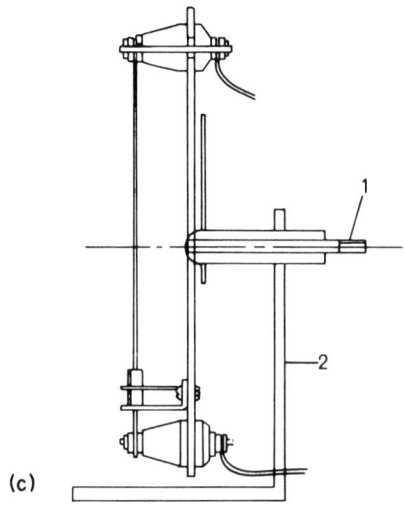

(c)

FIG. 1 (Continued)

<u>PIN AND CYLINDER TECHNIQUE</u>

Figure 2 illustrates the principle of the pin and cylinder technique. A molybdenum rod with a rounded tip rides on top of a short length of rotating molybdenum thin-wall tubing. The friction force is transmitted by means of a stiff linkage to one end of a flexible beam. A calibration of beam deflection versus force must be performed prior to the experiment. This may be done with a sensitive spring balance. As in the critical-angle method, deflection may be indicated by means of a pointer on the beam, or by the angular deflection of a reflected light beam. [For precise measurements of transient friction phenomena (e.g., stick-slip processes), strain gauges could be mounted on the beam and their output monitored on an oscilloscope.] The pin and cylinder technique is readily adaptable to studying the effect of sliding velocity on friction through the use of a variable speed drive motor.

The principal disadvantage of the pin and cylinder technique is the difficulty of assuring adequate heating of the sliding surfaces. A simple internal heater is shown in Fig. 2, although other methods such as electron bombardment of the slowly rotating cylinder, or induction heating may also be tried. The pin should preferably be lifted slightly off the cylinder during heating. This may be accomplished by magnetic coupling techniques, or by means of an additional motion feedthrough. To prevent annealing of the flexible beam, radiation shielding should be considered. Low sliding speeds (<1 rpm) will minimize vibration and bouncing of the pin.

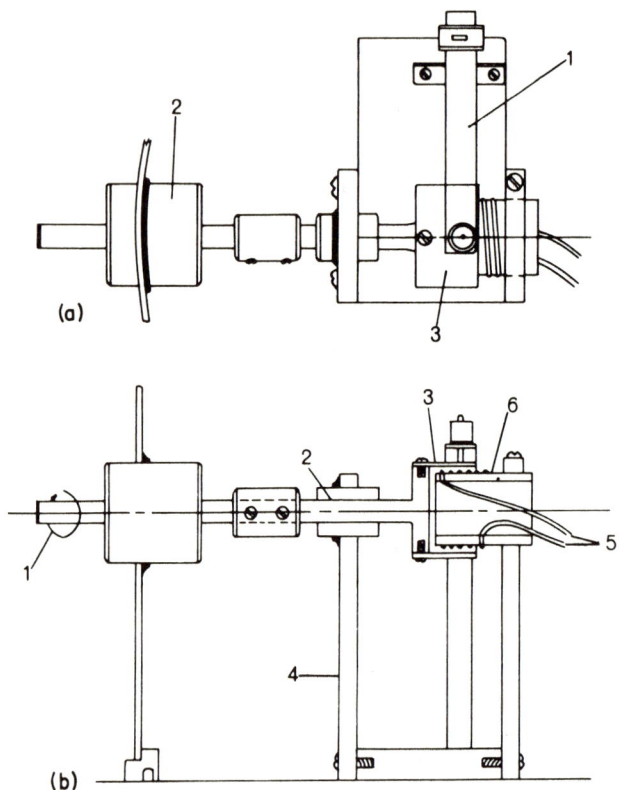

FIG. 2. Apparatus for the pin and cylinder method. (a) Top view. 1. 1/16"-stainless steel, 2. rotary motion feedthrough, 3. 1" diam-0.010 moly tube. (b) Side view. 1. Drive with variable speed motor, 2. Free fit, 3. 1" diam-0.010 moly tube, 4. Bearing support, 5. heater leads, 6. 3/4" O.D. aluminum oxide tube wound with tungsten wire. (c) End view. 1. light beam, 2. weight, 3. 1 mm diam moly rod, 4. mirror, 5. Pivot, 6. Rod lifter (optional), 7. 0.020 x 1/2" Spring steel, 8. Heat shield (optional).

6.2. STUDY OF FRICTION

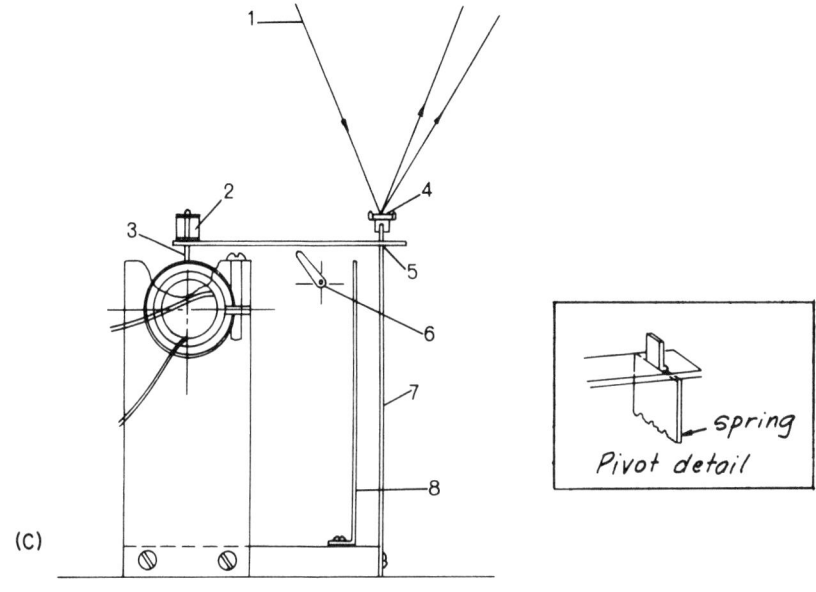

FIG. 2 (Continued)

BIBLIOGRAPHY

1. F.P. Bowden and D. Tabor, The Friction and Lubrication of Solids, Part 1, Oxford University Press, London, 1954.

2. I.V. Kragelskii, Friction and Wear (L. Ronson, trans.), Butterworth, Inc., Washington, D.C., 1965.

3. E. Rabinowicz, Friction and Wear of Materials, Wiley, New York, 1965.

4. E.R. Braithwaite, Solid Lubricants and Surfaces, Pergamon, New York, 1964.

Experiment 6.3

MEASUREMENT OF THE MEAN FREE PATH
OF CONDUCTION ELECTRONS IN SILVER

R. Olson and J. Wilson

Department of Physics
San Fernando Valley State College
Northridge, California

INTRODUCTION

Electrical conductivity depends on (among other things as well) the mean free path of the conduction electrons. A fairly simple and plausible form for the conductivity may be obtained from the Drude-Lorenz free electron theory [1,2]:

$$\sigma = \frac{ne^2 \ell}{2m\bar{v}} \qquad (1)$$

In this expression, σ is the conductivity, n is the number of conduction electrons per unit volume, e is the specific charge of an electron, m is the mass of the electron, $\bar{v}$ is the average randomly directed speed, and ℓ is the mean free path.

The mean free path, ℓ, appearing in the above expression, refers to the average distance traveled between collisions of an electron with a lattice atom. Collisions with the walls of the conducting body are neglected. Clearly, if collisions with the walls were taken into account, there would be an effective lowering of the value of ℓ, and hence a lowering of the conductivity. Thus, near the surface of a conductor (a depth less than the mean free path), the electrons suffer extra collisions, and their mean free path is less than that for electrons in the interior of the conductor. By preparing samples of conducting material with a transverse dimension comparable with the mean free path, the conductivity of the material is reduced.

Fuchs(3) in 1938 worked out the relation [Eq. (2)] between the measured conductivity of a thin film, the "bulk" conductivity, the mean free path, ℓ, and the sample thickness, a:

$$\frac{\sigma \text{ Film}}{\sigma \text{ Bulk}} = 1 - \frac{3\ell}{8a} + \frac{3\ell}{2a} \int_1^\infty \left(\frac{1}{x^3} - \frac{1}{x^5}\right) e^{\frac{-ax}{\ell}} dx \qquad (2)$$

The computed ratios σFilm/σBulk, and its reciprocals, are given in Table 1 and Fig. 1, as a function of a/ℓ. These values have been computed from Eq. (2).

TABLE 1

Conductivity Ratio as a Function of Film Thickness

$\frac{a}{\ell}$	$\frac{\sigma \text{ Film}}{\sigma \text{ Bulk}}$	$\frac{\sigma \text{ Bulk}}{\sigma \text{ Film}}$
0.001	0.00550	182
0.002	0.0100	100.4
0.005	0.0216	46.4
0.01	0.0378	26.5
0.02	0.0654	15.3
0.05	0.130	7.69
0.1	0.212	4.72
0.2	0.333	3.00
0.5	0.526	1.90
1	0.684	1.462
2	0.819	1.221
5	0.925	1.081
10	0.963	1.0390
20	0.9809	1.0191
50	0.9924	1.0076
100	0.9962	1.0038

6.3. MEAN FREE PATH OF CONDUCTION ELECTRONS

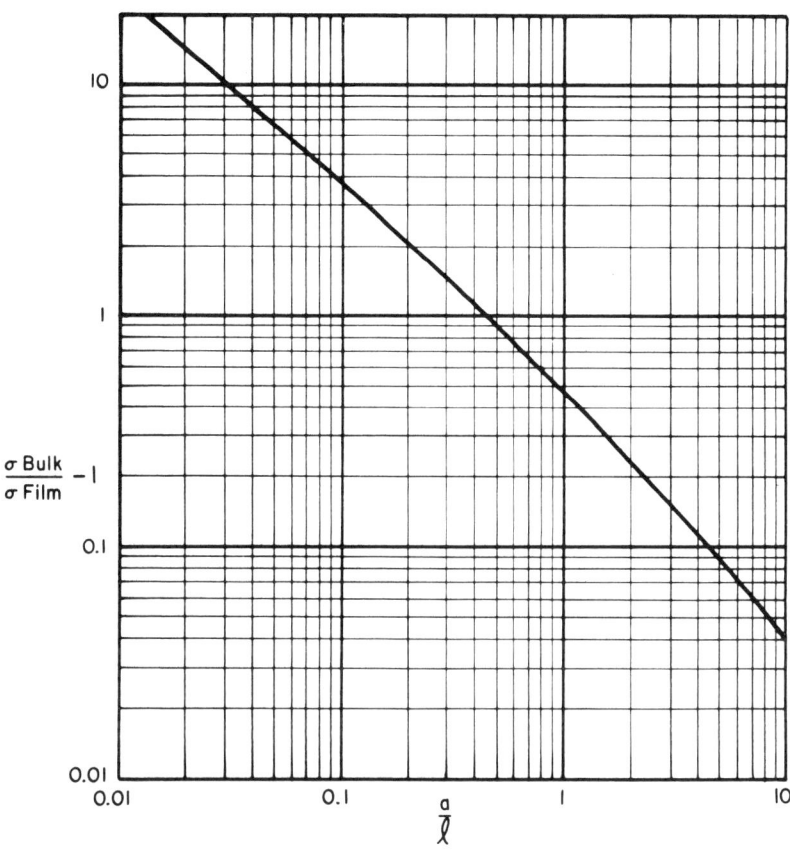

FIG. 1. Conductivity ratio as a function of film thickness.

EXPERIMENTAL SYSTEM

Equipment and supplies:
1. High vacuum system, capable of reaching 5×10^{-6} Torr
2. Evaporation source and power supply

    Example: Allen Jones type B-26b molybdenum or tantalum boat and 7 V 100 A transformer, with a Variac.

3. Glass bell jar at least 18 in. maximum dimensions, with safety shield

4. Ohmmeter capable of conveniently reading resistances between 1/2 and 5000 Ω

5. Optical bench with a photometer head--or a calibrated photometer

6. Comparator or traveling microscope

7. Light box with cover of ground glass or translucent plastic

8. Pellets of silver, 0.02 g to 0.2 g

9. Microscope slide holder for vacuum system

10. Shutter (optical) and control rod

EXPERIMENTAL PROCEDURE

A. PREPARATION OF SAMPLES

Two steps are necessary to provide samples for easy resistance measurement: (1) First, comparatively heavy but slightly "feather edged" deposits must be made near the ends of the substrate to provide relatively equipotential regions. (2) Then the substrate with the feather edge deposits are recoated with the desired range of thicknesses.

1. Eight to twenty clean flat glass substrates of convenient size (1/2 in. to 1 in. wide, 1 to 2 in. long) may be supported in a row across a flat bevelled band as shown in Fig. 2. The purpose of the bevel is to provide some feather edging of the deposit to ensure good contact with the film to be deposited later. The same material must be deposited in this step as that which will be studied later. This lessens the chance of error due to thermoelectric potentials incurred during the resistance measurements. For a spacing of 6 in. from evaporator to the substrates, about 0.2 g of silver will be sufficient.

2. Six to ten of the substrates prepared as in step 1 are then supported in a staircase fashion as shown in Fig. 3. The ratio of nearest to farthest distance from the evaporator

6.3. MEAN FREE PATH OF CONDUCTION ELECTRONS

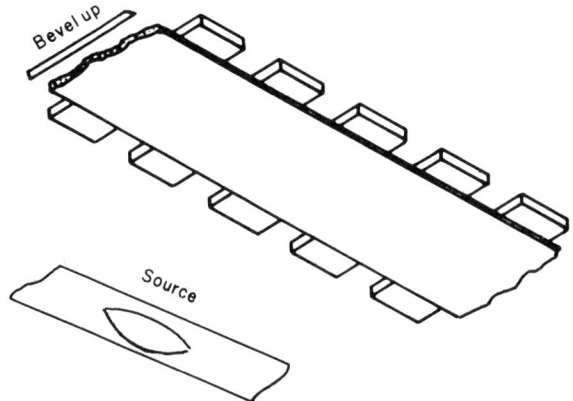

FIG. 2. Substrates arranged for end coating.

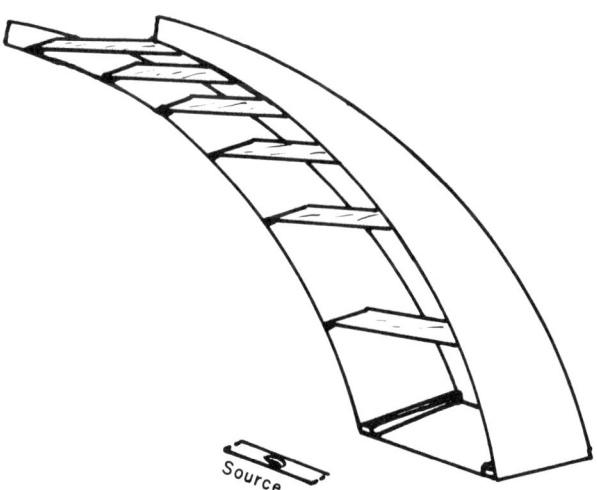

FIG. 3. Thin coating application arrangement.

source should be about four. It is suggested that for 1/2 in.-wide substrates the nearest distance be 4 in.

For larger substrates the distances will have to be proportionately increased.

The substrates are now ready for coating. To obtain consistent results and results essentially applicable to the values of Table 1, it is necessary that the background pressure be less than 2×10^{-5} Torr. Also, the material to be evaporated must be fairly well outgassed after melting. When the system pressure is acceptable, and the material molten, the evaporation may commence.

With no shutter

If no shutter is provided between the evaporant and substrates, all of the molten silver must be evaporated within a few seconds, by suddenly increasing the power to the evaporator. Some trial and error must generally be exercised to determine the best increase in power. Sixty to eighty per cent increase may be tried initially. Twenty to forty g of silver will be sufficient for coating specimens from 4 to 20 in. from the evaporant source. If the distance from the source to the nearest substrate is greater, the amount of silver necessary will have to be increased by the square of the ratio of this distance to 4 in.

With a shutter

The procedure to be followed is as described above, except that the power input is to be gradually increased after outgassing, with the shutter shadowing the substrates. When the glass bell jar shows a coating forming--this can be detected by looking at the heated boat--the power input may be increased about 30% and the shutter moved suddenly. When evaporation has ceased, the boat may be cooled. The coated substrates should now be allowed to remain under vacuum for a half hour before removal and inspection.

B. INSPECTION OF SAMPLES AND RESISTANCE MEASUREMENT

Great care should be exercised in removing the coated substrates from the vacuum system to the area where electrical measurements are to be made. Also, time wasting should be held to a minimum, although the resistance of the samples will not change much in ordinary atmospheric environments in the course of a few days. A simple visual inspection will tell whether useful samples have been obtained. If one <u>cannot see</u> through the coating, they are <u>too thick</u>. If one can see through them quite <u>clearly</u>, they are <u>too thin</u>. Several of the samples may be of acceptable thickness, and the resistance measured by an ohmmeter.

6.3. MEAN FREE PATH OF CONDUCTION ELECTRONS 187

Make several determinations of the resistance by applying the probes lightly to different locations on the heavily coated ends. If a reading variation is noted exceeding 5%, discard the sample.

After measuring resistance, the sample dimensions may be measured.

C. DETERMINATION OF SAMPLE DIMENSIONS

If care is exercised, the length and width of the thin coating may be determined by use of a vernier calipers with sufficient accuracy. However, a traveling microscope provides a better determination of dimensions and simultaneously allows an inspection of the film under a microscope. If any cracks are seen--especially perpendicular to the current flow--the sample should be discarded.

Optical attenuation by the coatings provides a simple means of thickness determination. Two convenient means are suggested:

1. Using an optical bench photometer head:
   In a darkened room, place on opposite ends of an optical bench two small flashlight bulbs. Find the location of the photometer head which provides equal light intensities and record the distance x_1 from one bulb to the center of the photometer. Then place the coated substrate in front of that side of the photometer head where x_1 was measured and move the bulb toward the photometer until equal intensities are again obtained. Record the new distance from the photometer center x_2. The percentage light transmission is then (assuming an inverse square law):

   $$\text{transmission} = 100 \frac{(x_2)^2}{(x_1)^2} \% \qquad (3)$$

   Next, for each coating's transmission value, read off the thickness from Fig. 4.

2. With a photoelectric cell, exposure meter, or similar device:
   Place one small bulb in a black box with a hole slightly smaller than the coated substrate. Mount the combination on an optical bench at one end with the aperture aligned down the bench. At the other end place a photoelectric cell and note its reading and the distance x_1 from the bulb. Then place the coating over the aperture and move the photocell until the same reading is obtained. Record the distance x_2 from the bulb. The percentage transmission may be calculated from Eq. (3) and the thickness read from Fig. 4.

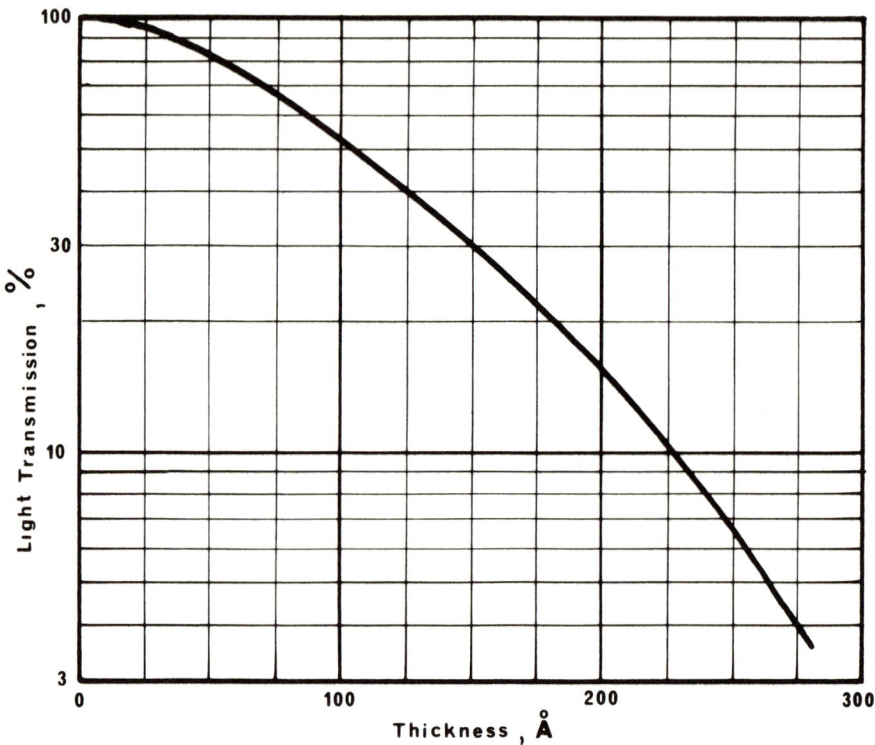

FIG. 4. Light transmission as a function of film thickness.

COMPUTATIONS

1. From the resistance and dimension measurements, calculate the conductivity from the following equation:

$$\sigma_{Film} = \frac{b}{Rac} \qquad (4)$$

where b is the length (parallel to the current), c is the width, a is the thickness, and R is the resistance.

2. Assuming the bulk conductivity of silver at 20°C as 1.63×10^{-8} (ohm/meter)$^{-1}$ calculate the quantity:

6.3. MEAN FREE PATH OF CONDUCTION ELECTRONS

$$\frac{\sigma_{Bulk}}{\sigma_{Film}} - 1$$

for each sample and tabulate the results.

3. For each sample, read off the value of a/ℓ from Fig. 1.

4. Calculate ℓ for each sample and find the average ℓ and its standard deviation. Compare your results with that of Reynolds and Stillwell[4] for silver:

$$\ell = 5.20 \times 10^{-7} \text{ m}$$

or $\ell = 520$ Å

REFERENCES

1. R. Resnick and D. Holliday, *Physics*, Part II, Wiley, New York, 1966, Chap. 31.

2. E. Sonheimer, *Advances in Physics*, 1, 1, (1952).

3. K. Fuchs, *Proc. Cambridge Phil. Soc.*, 34, 100 (1938).

4. F. Reynolds and G. Stillwell, *Phys. Rev.* 88, 418 (1952).

5. W. Bond, *Journal of the Optical Society of America*, 44, 435 (1954).

Editorial Comment (MTT). Before a student undertakes an experiment to measure the electron mean free path of a metal thin film, he should give thought to some of the complexities that can be encountered in the interpretation of the data obtained in such an experiment. These comments are not meant to dissuade the reader from doing such experiments, but on the contrary to keep him from becoming discouraged if nonreproducible or unrealistic data is obtained. In fact the prediction of thin film conductivity from known bulk properties and known film deposition parameters is one of the more difficult problems encountered in thin film investigations. Also, for many thin film applications, such as in the fabrication of thin film passive components, it is desirable to control the conductivity. At the present time neither of these goals have been achieved completely. It should be mentioned that remarkable progress has been made at controlling the film conductivity of such materials as tantalum using reactive sputtering for deposition.

Thin film conductivity is determined by the scattering of electrons from lattice thermal vibrations, the film walls, grain boundaries and other film growth defects, and the type and number of impurities incorporated in the film during its formation. All of these factors are sensitive to the condition under which the films are produced. These conditions can vary from one deposition to another even in the same vacuum chamber, making reproducible results difficult though not impossible to achieve. An estimate of the electron mean free path can be obtained by fitting the data to a particular theory, but the significance of this number is open to question without a thorough knowledge of the crystallinity, structure, texture, and impurity content of the film. The conduction mechanisms become even more complicated for very thin films (50 Å) which are composed of disconnected islands.

This problem is a challenging one and although it has been worked on for many years, it is still being actively pursued. The preceding experiment is a good introduction to the field. Some references in addition to those given in the experiment are included below.

BIBLIOGRAPHY

1. L. Holland, Vacuum Deposition of Thin Films, Wiley, New York, 1961.

2. H. Mayer, Structure and Properties of Thin Films, (C.A. Neugebauer, J.B. Newkirk, and D.A. Vermelyia, eds.), Wiley, New York 1959. (This paper includes a fairly extensive bibliography to earlier work.)

3. A.A. Milgram and Chih Shun Lu, J. Appl. Phys., 37. 4773 (1966).

4. E. Ditlefsen and J. Lothe, Phil. Mag. (G.B.), 14, 759 (1966).

5. H.J. Jurestschke, Surface Sci., 2, 40 (1964): Surface Sci. 5, 111 (1966); Surface Sci., 5, 171 (1966); J. Appl. Phys., 37 435 (1966).

6. W.F. Leonard and R.L. Ramsey, J. Appl. Phys., 37, 3634 (1966).

7. P.M. Tomchuk and R.D. Fedorovich, Soviet Physics Solid State, 8 (1967).

6.3. MEAN FREE PATH OF CONDUCTION ELECTRONS

8. T.J. Coutts and G.G. Mathews, Proc. Phys. Soc. (G.B.), 90, 1175 (1967).

9. I.O. Kulik, J.E.T.P. Letters, 5, 345 (1967).

10. Z.H. Meiksin and R.A. Hudzenski, J. Apply. Phys., 38, 4490 (1967).

11. L. Sander, J. Phys. Chem. Solids, 29, 291 (1968).

12. L.G. Aslamazov and A.I. Larkin, Phys. Letters, 26A, 238 (1968).

13. J. Feder and T. Jossang, Phys. Norveg, 1, 217 (1964).

14. M.S.P. Lucas, J. Appl. Phys., 38, 1632 (1965).

15. P. Jourdain and P. Thureau, C.R. Acad. Sci. (France), 27. 2015 (1963).

16. A.A. Hirsch and N. Friedman, Physica, 30. 389 (1964).

17. A. Gaide and P. Wyder, Thin Layers, Conference Liege, 411 (1961).

18. S. Offret and B. Vodor, Thin Layers, Conference Liege, 370, (1961).

19. A. Deveni and R. Meneile, Rev. de Physique (Roumania), 6, 491 (1961).

Experiment 6.4

CONSTRUCTION AND USE OF A CATHODE RAY TUBE

B.R.F. Kendall and H.M. Luther

Department of Physics and Ionosphere
Research Laboratory
Pennsylvania State University
University Park, Pennsylvania

INTRODUCTION

This project is intended to introduce students to the basic principles of producing, focusing, deflecting, and detecting electron beams in a vacuum environment. The same principles apply to a wide range of other electron-optical and ion-optical apparatus used in vacuum physics.

The project is intended to be carried out with the aid of a set of interchangeable electron-optical components which has been described in a recent publication(1). The same set of components makes a wide variety of other electron-optical and ion-optical devices. The published paper gives sufficient information for the construction of the necessary parts.

Figure 1(a) shows a simplified diagram of a typical electrostatic-deflection cathode ray tube, while Fig. 1(b) shows one of many arrangements of electrodes which can be used in such a tube to achieve the desired results.

In the electron gun, electrons are released by thermionic emission from a hot filament or cathode and accelerated along the axis of the device by the field between the filament and the accelerating electrode. The theory of thermionic emission is discussed in many textbooks (see, for example, Ref. 2).

The emerging electron beam from the electron gun is normally rather divergent. The beam must be focused. This is done by means

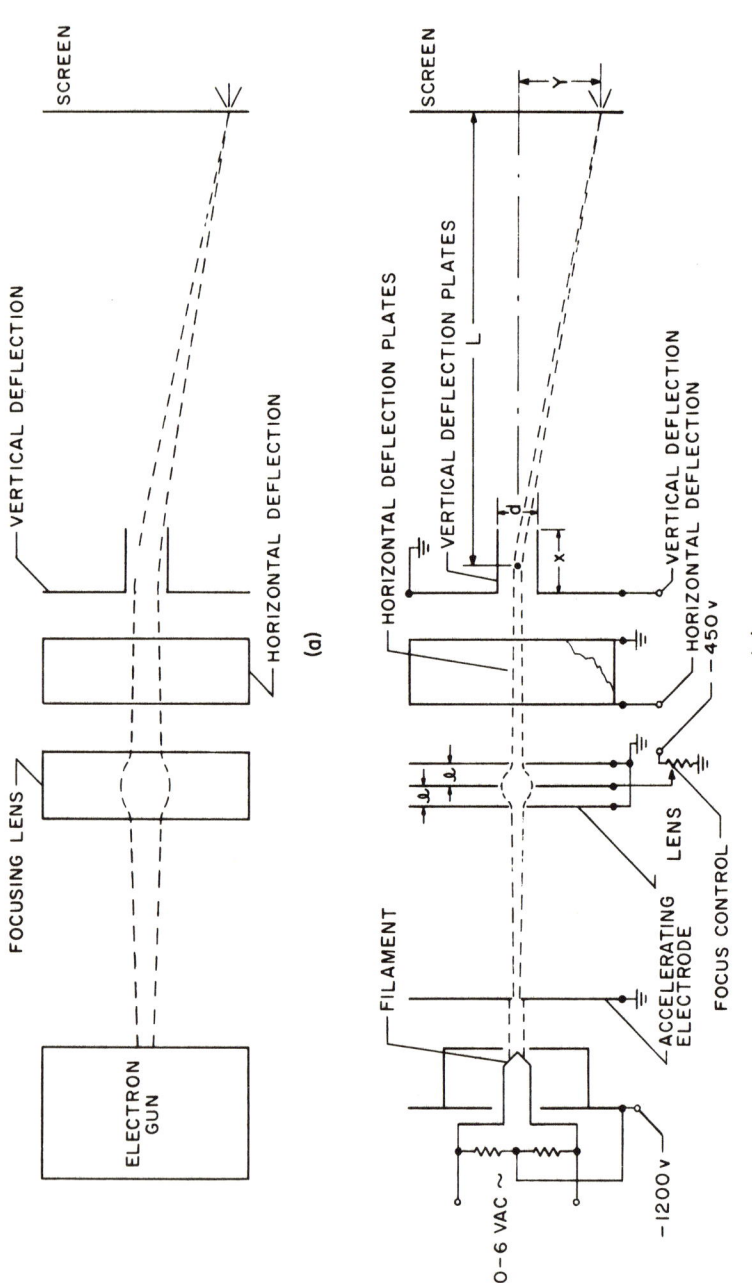

FIG. 1. Essentials of electrostatic-deflection cathode-ray tube. Reprinted from the American Journal of Physics with the permission of the editor.

6.4. CATHODE RAY TUBE

of an electron lens. The focusing action can be achieved by suitably shaped electrostatic or magnetic fields. In this experiment we use one of the simplest types of electron lenses, a symmetrical aperture lens having three electrodes. Focal length is varied by adjusting the potential of the center electrode. The relevant equation is:

$$F \simeq 2\ell \frac{V_o V_i}{(V_i - V_o)^2} \tag{1}$$

where F is the focal length measured from the center of the lens, ℓ is as shown in Fig. 1(b), and V_o and V_i are the potentials of the outer and inner electrodes, respectively, relative to the electron emitter. The equation holds approximately provided V_o/V_i is less than about 4. Commercial cathode ray tubes usually use cylindrical electrostatic lenses, but the basic principle is the same. The theory of these electrostatic lenses, and of magnetic focusing lenses, is given in the literature(3-4).

Deflection of the electron beam also can be achieved either electrostatically or magnetically. In this experiment we use two pairs of parallel plates to produce electric fields in the appropriate directions for deflecting the electron beam. The relevant equation is

$$Y = \frac{Lx}{2d} \frac{V_d}{V_o} \tag{2}$$

where Y, L, x, and d are as shown in Fig. 1(b), V_d is the potential difference between the vertical deflection plates, and V_o is the mean potential of the deflection plates relative to the electron emitter. Because the electron motion between the deflecting plates is parabolic, it can be shown that the central ray of the emerging electron beam behaves as if it had been deflected instantaneously at the center of the deflection system(2,3).

EXPERIMENTAL DESIGN

Figure 2 shows a working cathode ray tube assembled by a student from a set of standard components. The basic structure consists of a 0.030-in. thick stainless steel main frame A and sliding support frame B. They are linked by the four solid 1/8-in. diameter ceramic rods C.

These rods carry the fixed retaining clips D, a large number of 1/4-in.-o.d. ceramic spacers E, Inconel X compression springs F, and adjustable clamps G. Most of the electrodes, including the filament

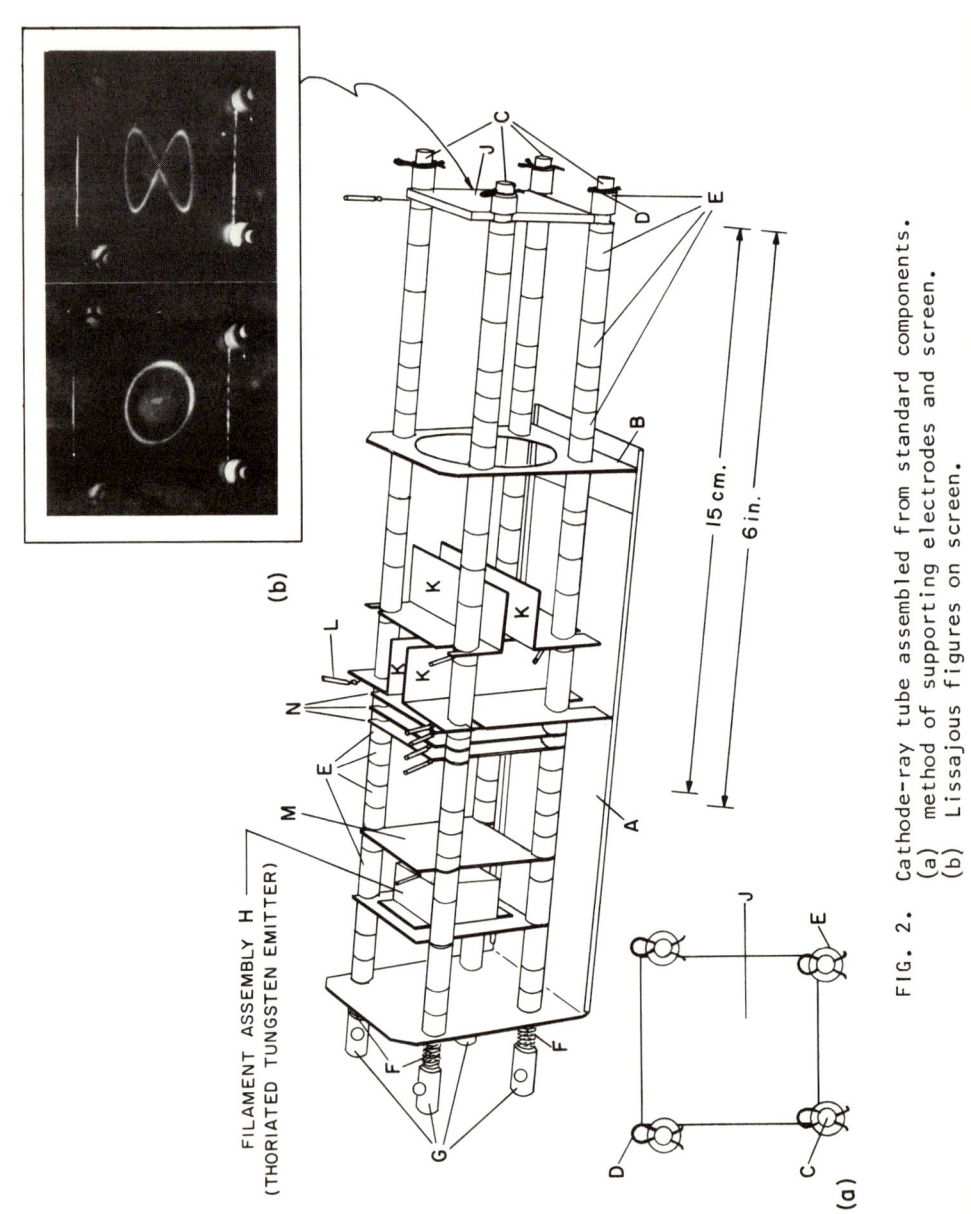

FIG. 2. Cathode-ray tube assembled from standard components. (a) method of supporting electrodes and screen. (b) Lissajous figures on screen.

6.4. CATHODE RAY TUBE

assembly H, are based on identical 1 1/2 in. square, 0.020-in. thick stainless steel plates. The 1 1/2-in. square glass screen J is coated with a P5 phosphor. These electrodes, the filament assembly, and the screen are supported by the ceramic rods and positioned axially by the spacers as shown in the detail drawing at the lower left in Fig. 2. The only exceptions to this method of support occur in the case of the horizontal and vertical deflection plates K, which are each supported by only two ceramic rods. Axial positioning for the deflection plates is achieved with spacers, as with the square electrodes. Electrical connections to all electrodes are made through the miniature plug-in connectors L.

This type of construction is self-aligning and the square plates are constrained in all but the vertical direction, in which movement is prevented by gravity and friction. Electrodes are easily added, removed, or shifted to new positions after compressing the appropriate springs to remove the frictional forces.

Decisions as to the precise positioning of electrodes are left up to the student. Suitable electrode positions should be determined theoretically and then tested in practice. Approximate electrode positions can be obtained, if desired, by scaling from Fig. 2. It should be stressed, however, that Fig. 2 shows only one of many possible configurations.

The potentials to be applied to the various electrodes are easily determined for a given geometry if the theory has been properly understood. The potentials shown in Fig. 1(b) were found satisfactory in the original project.

As an aid to understanding the motions of the electrons in this apparatus, it is useful to add two "phosphor-grids" to the basic structure. These are standard 1 1/2-in. square frames carrying high-transmission (90%) nickel meshes which have been coated with the same phosphor as was used for the screen. They indicate the cross section of the electron beam passing through them while having a negligible effect on the beam intensity. In this experiment one phosphor-grid should be placed between the electron gun and the focusing lens, and the other between the vertical deflection plates and the screen. Both should be grounded.

EXPERIMENTAL PROCEDURE

Assemble the cathode ray tube and check that all electrodes are secure and parallel to one another. Make sure that the circular apertures in electrodes H, M, and N (Fig. 2) are co-axial. Install the device in your vacuum chamber and connect up the electrical wiring. With the filament current turned to zero, check all potentials on the electrodes to make sure that they are correct. Remember that very high voltages are being applied.

Close the vacuum chamber and pump it down to a pressure below about 10^{-5} Torr. With zero-field conditions between the two pairs of deflecting plates, apply potentials to the appropriate electrodes and slowly raise the filament current. As soon as a light spot appears on the first phosphor-grid, you have adequate intensity. Now adjust the focusing potential to obtain the smallest light spot on the screen, J in Fig. 2. Notice the cross section of the electron beam on the phosphor-grid between the deflection plates and screen. It should indicate a linear beam convergence.

QUESTIONS

Center the beam on the screen by suitable adjustment of deflecting voltages. If substantial deflecting voltages are needed for centering, check for asymmetries in the structure. Is there any evidence of beam deflection by potentials on the wiring inside the vacuum chamber? If so, how might this be reduced? Is there any evidence of insulators or glass surfaces in the vacuum chamber charging up? How might this be avoided?

With the beam centered on the screen, observe the effect of magnetic fields at various points along the flight path of the electrons. First use a bar magnet. After you understand the relationship between field and deflection, try the effect of electrical meters such as those used in ion gauge controllers. Then try putting an electric motor, such as those used to drive vacuum pumps, in various positions near the apparatus while both motor and apparatus are running. Comment on your results. Are you sure that the performance of your cathode ray tube is not being affected by steady or fluctuating magnetic fields associated with your pumping system? If not, how might this problem be overcome?

After you are reasonably sure that your cathode-ray tube is operating as intended, measure the deflection sensitivities in cm/V at the screen for small vertical and horizontal deviations about the central position. Are the deflection sensitivities the same for both sets of deflection plates? Explain. Deflect the electron beam to a point near the edge of the screen

6.4. CATHODE RAY TUBE

and remeasure the deflection sensitivities. Are they different from your original readings? Can you think of reasons why this should be so?

The plates which are used for vertical deflections have elongated slots so that their spacing, d in Fig. 1(b), can be varied. Try altering their spacing and notice how the deflection sensitivity changes. Does it change according to theory?

Observe the effect of increasing the filament temperature and thus increasing the electron current (Note: in practical cathode-ray tubes this is done by means of a separate control electrode). What happens to the spot size? Return the electron current to its original value.

Progressively raise the pressure to about 5×10^{-3} Torr, preferably using an inert gas. Comment on any variations you notice in the size of the focused spot. Remembering that there will be ionication along the path of the beam, can you account for the initial reduction in spot diameter which is sometimes seen under these conditions? If ions are formed in a region where the electrons are being accelerated, can you predict their probable trajectories?

Use your cathode ray tube to display high-frequency as signals. Comment on its performance. Examples of Lissajous patterns obtained with the prototype are given at top right in Fig. 2.

A high gas pressure in the cathode ray tube also permits the path of the electron beam to be seen. Thus, using the magnet motor and other deflection techniques the path of the beam is made visible.

BIBLIOGRAPHY

1. B.R.F. Kendall and H.M. Luther, Am. J. Phys., 34, 580 (1966).

2. K.R. Spangenberg, Vacuum Tubes, McGraw-Hill, New York,

3. E.V. Cosslett, Electron Optics, 2nd edition, Clarendon Press, Oxford.

4. J.R. Pierce, Theory and Design of Electron Beams, D. Van Nostrand, Princeton, 1954.

5. F. Rosebury, Handbook of Electron Tube and Vacuum Techniques, Addison-Wesley, Reading, Mass., 1956.

Experiment 6.5

CONSTRUCTION OF A VACUUM TRIODE
USING SOLDER GLASS TECHNIQUES

J. G. King and J. Orsula

Massachusetts Institute of Technology
Cambridge, Massachusetts

INTRODUCTION

Solder Glass technique[1] offers an easy method of construction of all-glass vacuum envelopes. It is based on the use of commercial electron tube headers which are joined to sections of glass tubing, of suitable diameter and length, by the use of Solder Glass[2], as shown in Fig. 1. The headers provide a number of vacuum-tight leads which are used to support an electronic system inside the envelope and provide electrical connections to the various electrodes of that system. All glasses involved in the construction of vacuum envelopes by this technique are of the soft glass type, having a coefficient of thermal expansion, roughly, $85\text{-}95 \times 10^{-7}$ per degree Centigrade. (Even though Solder Glass for joining hard glass parts is available, no hard glass headers are, at present, obtainable at reasonable cost. For this and other reasons, construction of hard glass envelopes by this technique is not a practical proposition.)

Vacuum envelopes built by Solder Glass technique are suitable for continuously pumped systems but they are especially attractive for sealed-off systems, because they can be effectively outgassed by baking[3]. In contrast with other materials that can be used to join commercial tube headers to glass tubing (epoxies, silicon compounds, caulking compounds, etc.) Solder Glass (Corning 7570) is genuine glass that makes a vacuum-tight seal as good as the glass parts joined with it. A well fired Solder Glass joint can be baked under vacuum up to 300°C without harm.

A triode of suitable geometry, constructed by Solder Glass technique, is useful for the study of several aspects of high vacuum

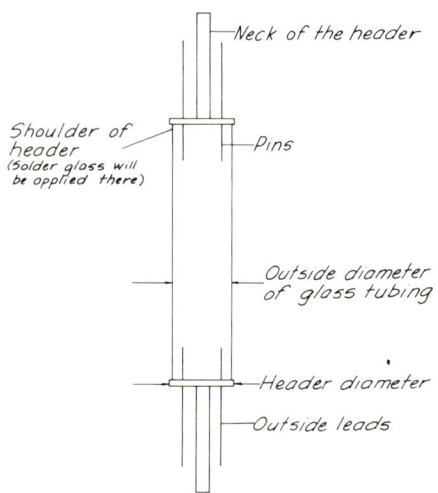

FIG. 1. Typical tube assembly using Solder Glass techniques.

work, especially the behavior of a sealed-off system using getters and ion pumping(4). It can be used as an ionization vacuum gauge(4) to monitor the pressure inside the envelope after the getters are activated. At the same time the changes in electron emission from the hot tungsten filament will illustrate the dependence of the efficiency of thermionic emission on the degree of clean-up within the envelope and the resulting surface condition of the emitter(5). When the vacuum gauge shows that the pressure has reached an equilibrium, the triode can be connected in a circuit suitable for measuring its electrical characteristics(6). It can also be connected in a practical circuit to show that it will function as a triode.

TUBE CONSTRUCTION PROCEDURE

A triode with planar grid and anode is easier to construct than the more usual, concentric, cylindrical system and the following description gives some details of the construction of such triode.

6.5. CONSTRUCTION OF A VACUUM TRIODE

The electrodes are mounted on the pins of a 14-pin header (RCA Type FSB 618 AB)*; the other header will carry the getters. The most suitable standard size of soft glass tubing to be used with these headers has 38 mm o.d. and about 35 mm i.d. The triode must be designed and built to fit inside this tubing.

Before construction of the system begins, the header should be inspected for possible cracks in the glass or other damage. If it is all right, its pins should be straightened into a symmetrical array, so that one can achieve symmetry and good alignment of all the electrodes that will be mounted on the header.†

As a general rule, all electrodes should be mounted at least 1 in. above the header in order to avoid having them excessively oxidized when the header joint is fired. This means that electrodes should be attached to the tops of the pins or, better still, that the electrodes have supporting wires attached to them and these, in turn, are connected to the pins. All metal-to-metal joints are made by electrical spot welding (7).

The anode is made of a rectangular piece of stainless steel sheet 0.005 in. thick, bent into a shallow rectangular channel, as shown in Fig. 2. A simple wire frame, made of stainless steel wire, 0.035 in. diameter, is attached to it and connects it to the pins of the header. The diameter of the wire, 0.035 in., was chosen in order to facilitate the spot welding of the frame to the pins of the header which are made of nickel.

The grid is made by spot welding straight, short pieces of 0.10 in. diameter tungsten wire across a frame like the one used for the support of the anode. It is made of the same stainless steel wire. The spacing of the tungsten wires will determine, besides other factors, the electrical characteristics of the triode. Spacing the wires from 0.125 to 0.250 in. is suggested.

---

*All materials and most of the equipment used in Solder Glass technique are standard commercial products. It is expected that all will be available shortly from a single source: Macalaster Scientific Co., Division of Raytheon Co., Route 111 and Everett Turnpike, Nashua, New Hampshire 03060.

†A 16-mm movie showing the construction of vacuum tubes by Solder Glass technique was produced for the College Physics Film Program by the Education Development Center, Inc., 55 Chapel Street, Newton, Mass. 02158 under a grant from National Science Foundation. It is available (Cat. No. 0468) from Modern Learning Aids, 1168 Commonwealth Avenue, Brighton, Mass.

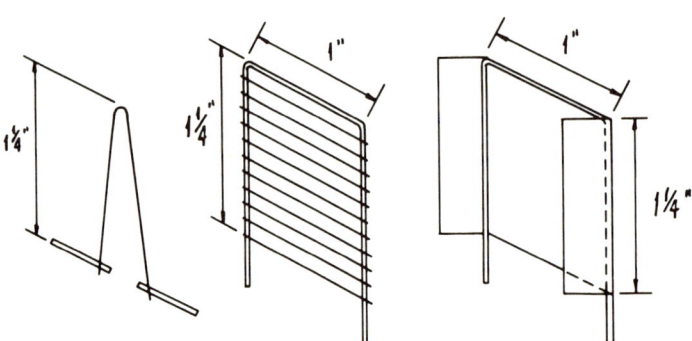

FIG. 2. An exploded view of the triode.

The filament is made of annealed tungsten wire, 0.005 in. diameter. Its total length should not exceed 2.5 in. If it is longer, the hairpin loop, into which it is formed, is not self-supporting when operated at high temperature in the horizontal position. Even if the filament is not longer than recommended, it is advisable to operate it only in the vertical position, preferably hanging down.

The spacing of the electrodes will also affect the electrical characteristics of the tube. Suggested distance between the planes of the filament and the grid is 0.125 in. and between grid and anode, 0.500 in.

The triode mounted on the header is shown in Fig. 3. In order to reduce possible effects of external fields and electrical charges on the glass tubing on the fields that we expect to set up between the electrodes, the anode has been shaped as shown. To reduce these effects even more, the triode will be surrounded by a cylindrical sleeve made of stainless steel wire cloth (mesh #100). This sleeve can be made to fit snugly into the glass tubing, so it will stay in place without being fastened to the glass. A spring, made of tungsten wire, 0.010 in. diameter, attached to a set of pins on the header behind the anode, will provide an electrical connection to the sleeve when the tubing is joined to the header.

The completed header should be cleaned and attached to the glass tubing. A simple cleaning procedure consists of a wash in a strong grease solvent such as trichlorethylene, a rinse in acetone, and final rinse in alcohol. A more thorough cleaning procedure should be adopted if possible.

The length of tubing that will be attached to the header should be about 2 in. longer than the sum of the lengths of the systems mounted on each header--in the case at hand about 6 in. of the total

6.5. CONSTRUCTION OF A VACUUM TRIODE

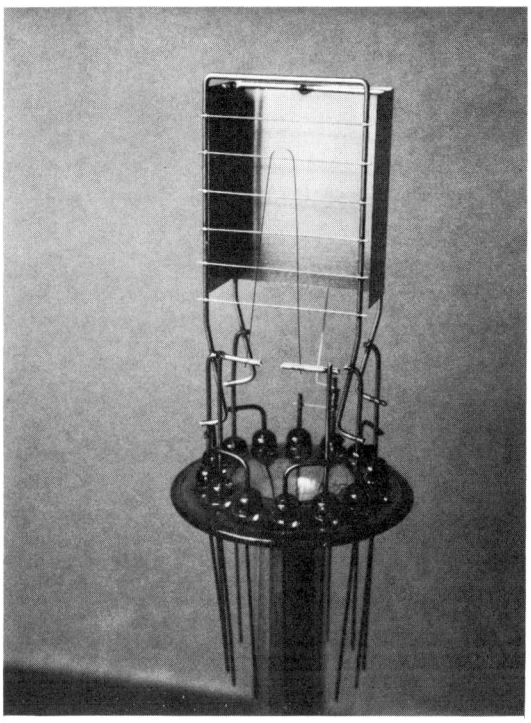

FIG. 3. The triode mounted on the header.

length of tubing comes out to 8 in. The tubing should be cut to the desired length by some method capable of producing a smooth cut. Uneven or ragged edges must be avoided.

The end of the tubing that is to be attached to the header is dipped into the Solder Glass paste about 1/4 in. deep, withdrawn slowly and tipped into horizontal position. It is then rotated slowly around the axis of the tubing until the paste produces an even coating and dries. A small electric fan will shorten the drying time to 2 or 3 min. The paste is then applied, with a small spatula, all around the shoulder of the header and, while it is still wet, the coated end of the tubing is lowered carefully over the triode onto the shoulder of the header; the assembly is left to dry undisturbed for 20 or 30 min. Again, a small fan or drying oven helps to cut down the waiting period. When the paste is dry, the joint can be handled gently and corrections made by adding a little paste, where needed. A joint made as described above is shown in Fig. 4.

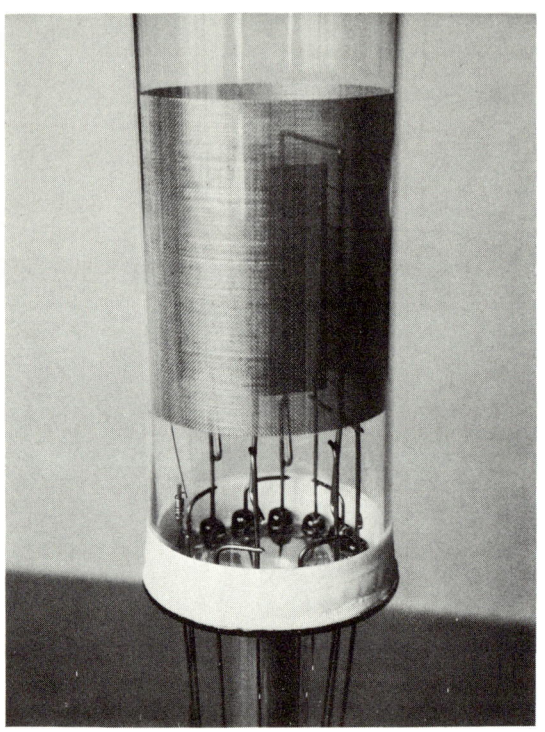

FIG. 4. Solder Glass joint ready for firing.

The joint is then fired in an electric ringheater. Firebricks with suitable size holes in them enclose the bottom and the top of the ringheater as shown in Figs. 5, 6, and 7. The clamp around the neck of the header at the top helps to prevent sagging of the header, should it get overheated. In order to fuse the Solder Glass powder, the joint has to be heated to about 550°C. Instead of measuring its temperature during the process, which is, to say the least, difficult, the time necessary to produce a good joint is found by experimenting with a few dummy joints. The ringheater shown in Fig. 6 is operated directly from an a.c. line and it takes about 12 min. to produce a well fired joint with it. Needless to say, the way the joint is enclosed in the ringheater must always be the same, if consistent results are expected. When the ringheater is turned off, the whole assembly is left undisturbed for 15 to 20 min and allowed to cool down slowly and evenly. This is essential!

6.5. CONSTRUCTION OF A VACUUM TRIODE

FIG. 5. Setting up to fire the joint.

All the waiting periods mentioned above can be used to start the construction of the getter header. It is suggested that 2 or 3 barium getters and 1 titanium getter be used. The getter should be oriented so that the deposits, which will be produced on the inside of the envelope, will not cover the same area. A header with two barium and one titanium getter is shown in Fig. 8. The circular disk, mounted on a wire frame, prevents getter deposits on the electrodes of the triode and, possibly, on the triode header, where they could do a lot of harm.

The cleaned getter header is then attached to the other end of the tubing as described above. When this joint is dry, a small drop of Solder Glass paste is put inside the neck of each header, as near as possible in the middle, and smeared around. This will help to produce vacuum-tight seals when the tube is pumped out and the necks

FIG. 6. Setting up to fire the joint; oven in place.

sealed off. The firing of the getter header joint completes the tube. All the outside leads have become oxidized and should be scraped clean with a piece of coarse emery cloth. Leads from paralleled pins should be twisted together and soldered. The tube is then ready for pumping.

The choice of a suitable pumping procedure will depend on equipment available and the amount of effort that the experimenter wants to spend, as well as on the planned use for the tube. Its most interesting use is as an ionization vacuum gauge to study the action of the getters and ion pumping in a sealed-off tube. For that purpose a simple procedure that will give satisfactory results with a minimum of effort is to pump the tube on a two-stage rotary vacuum pump, capable of pumping down to below 10^{-2} Torr, outgassing as many electrodes as possible by available means and then baking the whole tube at about 200°C for a few hours before it is finally sealed off.

The pump should be equipped with a vacuum gauge which covers the operating range of the pump. The tube can be attached to the pump by

6.5. CONSTRUCTION OF A VACUUM TRIODE

FIG. 7. Setting up to fire the joint; oven complete.

a short piece of vacuum quality rubber tubing. The other end of the tube is plugged up by a rubber stopper and the pump started. As soon as there is indication that the system works all right, a small electric oven is put around the neck of the tube that is plugged up by the rubber stopper. It should be centered about the ring of Solder Glass paste inside the neck and turned on. This oven, further called seal-off oven, shown in Figs. 9 and 10, is made of two parts hinged together, which makes it easy to put it in place and take it off. It is constructed, partly, of Pyrex glass tubing so one can see into it and observe how and when the seal forms. When the section of the neck inside the oven gets hot above its softening point, the atmospheric pressure makes it collapse and seal off. Without the Solder Glass on the inside, however, the seal may not be vacuum-tight, so it is important to put it there. When the collapsed part of the neck is about 1/2-in. long, the oven is turned off but left in place undisturbed for some 10-15 min, in order to allow the seal to cool down slowly and evenly.

FIG. 8. The getters mounted to the header.

With the far end sealed off, the pressure in the tube should now be below 10^{-2} Torr. When the seal-off oven is removed, the tube should be tested for leaks and if none are found, one can proceed to outgas the electrodes. The barium and titanium getters and the filament of the triode can be outgassed by heating to dull red (as observed in subdued roomlight) for at least 1 min each. A filament transformer (2.5 V secondary) connected to a Variac is all that is needed. The grid and anode cannot be outgassed by such a simple procedure. In order to outgas the glass envelope, a baking oven is built around the tube, using firebricks and fiberglass insulation. An attempt to illustrate the set-up is shown in Fig. 10, the baking oven partly open to reveal the tube inside. The temperature should be increased slowly to about 200°C, kept at that level for one or two hours, then decreased to about 150°C for another hour and finally sealed off using the same seal-off oven and procedure as before. The heater of the baking oven can be turned off a while before the seal-off oven is turned on. When the tube is sealed, the baking oven can be partly opened to allow the tube to cool down. In 15 min the tube should be cool enough to be taken off the pump and prepared for use.

6.5. CONSTRUCTION OF A VACUUM TRIODE

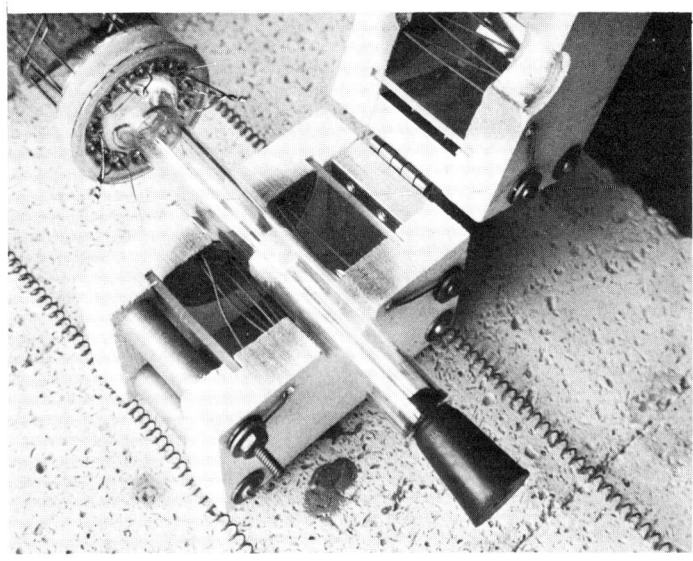

FIG. 9. The seal-off oven being placed around header neck.

FIG. 10. Partly assembled baking oven and seal-off oven in position for final sealing off.

The preparation consists of connecting about a 2-ft piece of insulated, stranded hook-up wire to each electrode of the triode, including the cylindrical shield. These connections are best made by soldering the wires to the leads. Slip short pieces of plastic sleeving over the soldered joints to prevent short circuits. Taping the wires to the neck of the header will prevent accidental breaking of leads, which are somewhat brittle at the point where they go through the glass.

EXPERIMENTAL PROCEDURE

The activation of getters comes first: Barium getters should be activated before the titanium. The getters are heated to about 1100°C (bright orange) and in a few seconds a deposit of barium should appear on the glass. One can produce either a thin (transparent) deposit, or let all the available barium be evaporated. The thin deposit offers a controllable way of using the available barium, and its behavior gives a qualitative indication of the pressure in the tube: When the deposit becomes saturated with gas, it becomes transparent. Thus, when the first layer of barium disappears, one can produce additional layers and, observing the rates at which these successive deposits get saturated, one can get some idea of how the clean-up proceeds.

All this time, of course, the triode should be connected to power supplies and turned on--the gases must be driven off the electrodes before they can be absorbed by the getters.

Since the triode can be used as an ionization vacuum gauge, a much more quantitative measure of the clean-up is possible with this tube. For initial rough work there is no need to calibrate the gauge, one can simply assume that the calibration constant S of our gauge is not too far from that of commercial products, about 10 A/Torr A. S is defined by:

$$ S = \frac{1}{p} \frac{I_{(+)}}{I_{(-)}} $$

where p is the pressure in Torr, $I_{(+)}$ is the ion current, and $I_{(-)}$ is the electron current. Both currents are measured in the same units. A typical range of $I_{(-)}$ is from 10^{-2} A to 10^{-4} A; 10^{-3} A is a good average to start with. From that it is seen that $I_{(+)}$ is of the order of 10^{-8} A at a pressure 10^{-6} Torr. Thus, one is faced with measuring small current for which, at present, a vacuum tube or solid state electrometer is the best choice.

6.5. CONSTRUCTION OF A VACUUM TRIODE

A typical triode, built as described above and connected as an ionization vacuum gauge, will indicate pressure in the range of 10^{-4} Torr (assuming $S = 10$) when it is turned on after the barium getters have been fired. If enough barium is available, the clean-up will decrease the pressure by about two orders of magnitude. The process may take from one to several days, depending mainly on the rate at which the parts outgas and on the amount of barium available. The outgassing of the grid and anode of the triode can be speeded up if they are connected to the positive terminal of a dc power supply capable of supplying about 50 mA at 500 V. The negative terminal is connected through a milliammeter to the center-tap of the filament transformer. The filament current is then turned up until the emission current reaches 20-30 mA. It is kept at that level for several minutes, while the appearance of the barium getter deposit is carefully watched. This is called outgassing by electron bombardment. If too much gas develops, an electrical discharge may take place inside the tube and damage the milliammeter as well as the power supply. To prevent it, the power supply should have a 50 mA fuse in its output. After a few minutes the emission current should be reduced to about 10 mA and the outgassing continued at this setting for about one hour. The triode can then be connected as an ionization vacuum gauge again.

If the barium deposit retains its visual appearance for several hours or longer, one can assume that the first step in clean-up has been accomplished. The titanium getter can now be activated. The pressure, as monitored by the gauge, will go up at first (outgassing) but should drop below the level obtained by barium alone. Again, this will take time, like the clean-up due to barium, but the gauge readings can come down another two orders of magnitude. A pressure of the order of 10^{-7} or 10^{-8} Torr may not seem spectacular by present day standards but it is adequate for a lot of interesting work, especially if it is realized that a large part of this residual pressure is due to noble gases which are not affected by the getters.

In addition to the behavior of the barium getter deposit, in relation to the pressure inside the tube, the behavior of the tungsten filament offers another indirect way of judging the pressure conditions in the tube. Although a heated tungsten filament will emit electrons even under fairly adverse conditions, under which other types of emitters would fail, the efficiency of the emission will be affected by the processes of gettering and ion pumping. Thus, measuring the filament current (all other things being equal) at various stages of the clean-up will furnish interesting insight into the behavior of the tungsten filament. If one thinks of the drastic treatment that tungsten metal has to undergo in order to yield thin wire, one will expect that on heating this wire to high temperature under vacuum, a considerable structural change must take place--and it is only fair to expect that this change will affect the electron emission from its surface.

BIBLIOGRAPHY

1. J.H. Owen Harries, Am. J. Phys., 28 (8), (1960).

2. R.H. Dalton, Am. J. Phys., 32 (6), (1964).

3. B.J. Todd, J. Appl. Phys., 26, (1955).

4. S. Dushman, Scientific Foundations of Vacuum Technique (J.M. Lafferty, ed.), 2nd edition, Wiley, New York, 1962.

5. W.B. Nottingham, MIT RLE Techn. Report 321, (Dec. 1956), or Handbuch der Physik, XXI, (1956).

6. H.V. Malmstadt and C.G. Enke, Electronics for Scientists, Benjamin, New York, Vol. 3, 1963.

7. J.G. King, Am. J. Phys., 32 (1), (1964).

Experiment 6.6

EXPERIMENTS USING SOLDER GLASS TECHNIQUES

D. Whitcomb

Motorola, Inc.
Phoenix, Arizona

INTRODUCTION

Many experiments in modern atomic physics require the use of high vacuum. The normal techniques for producing high vacuum require expensive and specialized equipment, keeping it out of the high school and small college laboratory. However, in 1959, Harries described a technique by which students could economically construct workable vacuum tubes (1,2). This technique uses simple but effective methods, and yet enables pressures as low as 10^{-7} or 10^{-8} Torr to be produced and maintained. Glass blowing is eliminated and only the simplest of joining and sealing operations are required. Normal hand tools are adequate for the mechanical assembly operations. Several experiments have since been described in the literature using Solder Glass techniques (3-5).

EXPERIMENTAL SYSTEM AND PROCEDURE

To form the envelope, glass tubing and commercially available multiwire glass stems are joined using Corning Solder Glass 7570. The solder glass is fused in an oven at 570°C, and the joints can be patched and reheated, and unsoldered and remade repeatedly.

A simple homemade oven is about 3 1/2 in. in diameter and about 3 in. long, using a nichrome heating element of about 330 W. A larger bakeout oven is used for 350°C bakeout of the components.

The internal hardware is made of 305 stainless steel, and is attached by crimping or simple screw clamps.

A cold cathode type of gauge is used to monitor the pressure and to pump the tube into the high vacuum region. Barium-aluminum and titanium getters are mounted in the envelope and are fired by resistance heating through the multiwire stem.

A typical tube is shown in Fig. 1. In the cathode ray tube, Willemite is used to produce a visible electron beam spot. Thoriated tungsten filamentary cathodes are used, although it is possible to process oxide cathodes by the procedure. The electrodes and glass parts are cleaned by washing in a household detergent solution and drying with warm air.

One method of exhausting the tube is to use a glass appendix filled with 6-14 mesh activated charcoal. The vacuum envelope, appendix, and the gauge are baked in the homemade oven at about 350°C for at least one hour while carbon dioxide from dry ice flows through the assembly. While heated, the input and output of the assembly are sealed off. The assembly is cooled, and the appendix containing the charcoal is further cooled by immersing in dry ice. The charcoal absorbs the gases and the pressure in the tube falls roughly to a value corresponding to that obtained with a commercial roughing pump(6). The appendix containing the charcoal is then sealed off and detached from the vacuum tube.

The Penning vacuum gauge is then switched on, and the barium-aluminum and the titanium getters are fired in that order. The characteristic blue glow of the carbon dioxide gas is seen before the getters are fired. It is important that the BaAl getter be fired first, since it has the property of removing the mixture of gases present in the "dirty" tube. Pressures in the order of 10^{-6} Torr are obtained after flashing the BaAl getter. After the Ti getter is fired, the remaining gases are pumped by the titanium film, and the pressure drops to about 10^{-8} Torr. The pumping action of the Penning-type vacuum gauge is adequate to maintain this low pressure, although most tubes will hold about 10^{-5} Torr without the pumping action of the gauge.

Instead of using the activated charcoal and carbon dioxide, a conventional mechanical pump can be used, but the tube should be baked out at 350°C before being sealed off from the mechanical pump.

COMMENTS

In the construction of the tubes, the effect of cleanliness on the pressures attained is clearly demonstrated.

6.6. EXPERIMENTS USING SOLDER GLASS TECHNIQUE

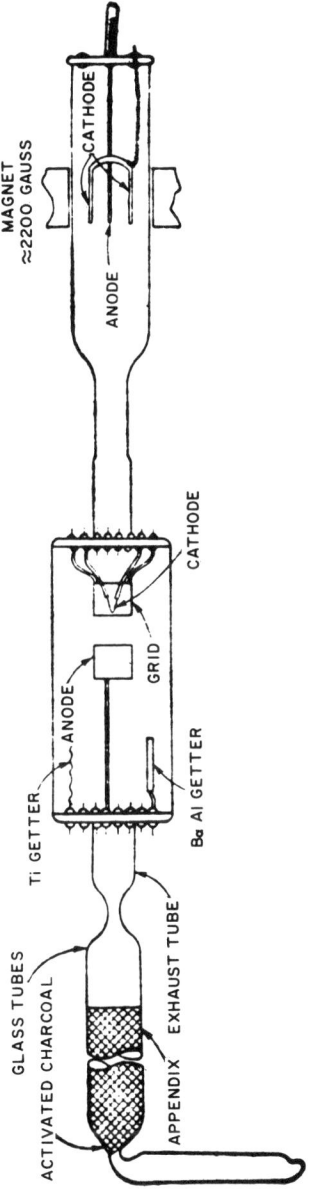

FIG. 1. Electron tube assembly.

The ability of the Penning-type gauge to pump gases and the properties of the BaAl and Ti getters are illustrated.

Experiments can be performed showing the effect of firing the getters in different orders.

Once completed, the electron tubes can be used to demonstrate the laws of electron optics.

Simple diodes, triodes, or a cathode ray tube can easily be constructed using simple parts.

This type of experiment is appropriate for a high school physics laboratory, yet teaches many of the basic elements of vacuum technology.

REFERENCES

1. J.H. Owen Harries, Electronics, 78 (June, 1959).

2. J.H. Owen Harries, MIT Physical Electronics Conference, 1960.

3. J.G. King, Am. J. Phys., 32(6), 483 (1964).

4. J. Rosenfeld and C. Tyler, Am. J. Phys., 33(10), 849 (1965).

5. J.G. King, Am. J. Phys., 32(6), 473 (1964).

6. S. Dushman, The Production and Measurement of High Vacua, General Electric Company, Schenectady, New York (1962).

Editorial Comment(VJH). For complete details of the Solder Glass technique, see Experiment 6.5.

Section 7

SPECULATIONS

7.1

ORIGINAL THOUGHT EXPERIMENTS

M. R. Carbone

Mason-Renshaw Industries
Carpinteria, California

Set up experiments related to the following statements and questions and make meaningful comments thereon.

1. Would it be possible and/or practical to put an electrical charge on the vapor stream in a diffusion pump and direct the stream by electric fields?

2. Would it be possible and/or practical to make a high vacuum pump by connecting a number of water aspirators in series? What if you replaced the water with diffusion pump oil (recirculating)?

3. A diffusion pump "works" because of the kinetic energy in the vapor stream, which is obtained from the pump's heater. Could the heater be replaced by a pressure pump which would put energy into the system by compressing the diffusion pump fluid?

   In a system of this type, what kind of discharge nozzle would work best? What about passing the fluid through an ultrasonic "whistle"?

4. Would it be possible or practical to pump a vacuum system by backfilling it with oxygen and then heating an iron filament by passing a current through it so that the iron would combine with the oxygen and lock it into iron oxide?

5. Is it possible or practical to create a standard vacuum for vacuum gauge calibration by filling a known volume with a pure condensable gas and then cooling the container to various cryogenic temperatures?

6. One simple vacuum experiment is the ringing-bell in the bell jar. As the chamber is pumped down, we no longer hear the bell. This may be considered a vacuum gauge with a transducer (the bell) and a receiver (our ear) coupled by the air in the chamber.

    How sensitive could this gauging system be made if we were to use a well coupled transducer and a receiver? What would be the advantages and disadvantages of a gauging system of this type?

7. Why not use a Tesla (high frequency spark coil) as an ion producing source in a vacuum gauge? If this is not practical by itself, how about using it <u>with</u> a hot filament or cold cathode? How about using a Tesla coil to generate ions in a titanium pump?

8. Why not use an RF frequency generator as an ion producing source in a vacuum gauge? If this is not practical by itself, how about using it with a hot filament or cold cathode? How about using an RF frequency generator to generate ions in a titanium pump?

9. Why not use a laser beam as an ion producing source in a vacuum gauge? If this is not practical by itself, how about using it with a hot filament or cold cathode? How about using a laser beam to generate ions in a titanium pump?

10. In thin film vacuum evaporation the metal being deposited is vaporized by heating. The metal molecules then travel in straight lines to hit the target. Would it be possible to ionize the metal molecules and then focus and direct them with appropriate circuits so as to precisely control the deposition pattern?

11. The Coanda effect is related to the flow of liquids and gases through specifically shaped orifices and across specifically shaped surfaces. The gases and liquids adhere tightly to certain shapes and reach very high velocities. Would it be possible to utilize this effect in the construction of diffusion pumps?

12. Contrary to common belief, noncharged particles <u>can</u> be moved by an electric field. All that is <u>necessary</u> is that the field be converging. If certain noncharged particles are placed in a converging field, the balanced charges on the particle will align themselves with the field. Since the field on the converging side of the particles is stronger than on the diverging side, the <u>forces</u> on the particle are not balanced and the particle <u>will</u> move in the direction of the converging field. This technique has

7.1 ORIGINAL THOUGHT EXPERIMENTS

been used to sort powders. Can this phenomena be used to direct the flow of gases in a vacuum? Can noncharged gases be thus directed to the hot wire of a thermocouple gauge to increase the gauge sensitivity?

13. Can a small tuning fork, or a large one for that matter, be used as a monitor for vacuum deposited thin films (like crystals are used now)?

14. Crystal oscillators used for the measurement of vacuum deposited thin films are severely sensitive to the temperature of their surface and must often be mounted on water cooled blocks. Could one mount the crystal on a closed container partially filled with an appropriate volatile liquid? As the temperature rose, would the liquid vaporize, carrying the heat quickly to the entire interior surface of the container, where it would be conducted to the exterior of the container and radiated away? In effect, we would have a small heat pipe acting as the heat sink.

15. Gases have different ionization potentials. Is it possible to construct a worthwhile gas analyzer utilizing this fact?

16. Would it be possible to use a crystal oscillator as a dew point indicator? Cooling the crystal would result in condensation on the crystal and a frequency shift.

17. Would it be practical to use a radioactive source as the ionizing element in a vacuum gauge?

18. We are all familiar with the creation of a vacuum when a steam filled container is cooled. Can this technique be used as a worthwhile pumping technique? Consider various materials and cooling methods. What about spraying a mist of a suitable liquid against a cryogenic surface?

19. Can the sensitivity of a thermocouple (or thermistor) gauge be improved by ionizing gas near the thermocouple and putting a negative voltage on the thermocouple to attract the ions? Thermocouples now rely on random impingement.

7.2

PROVOCATIVE IDEAS AND QUESTIONS

N. Milleron

Lawrence Radiation Laboratory
Berkeley, California

1. Pumping by "action at a distance"

Imagine a laser beam passing through a vacuum system. Discuss the probability of producing ionization, in the gas and on surfaces, by the laser action. Discuss the cross sections for ionization as a function of the photon energy and the electric field strength associated with the laser beam. Is it possible to ionize molecules with photons whose individual energy is less than the ionization potential of the particular molecular species? Look up multiphoton absorption effects. Can the ions thus produced be "sucked out" by appropriate electric fields? What would you do with the ions once the electric fields had operated on them? How could the ions be transferred into air?

2. Achieving light pipe free molecular flow conductance

Is it possible to change the surface reflection of molecules to achieve 100% transmission from entrance to exit of a pipe? How smooth would surfaces have to be? Can the required smoothness be accomplished mechanically? Discuss grazing angle encounters of molecules with the wall. What would happen if the light pipe were moving, for example, in outer space with a velocity high compared with the molecular velocity?

3. Washing off molecules bound to walls by chemical reaction

Would it be possible to clean up wall surfaces in a vacuum system by admitting a special chemical in liquid or gaseous form? Discuss the pros and cons of this notion. What magnitude of chemical reaction would be required to remove water in this manner?

4. <u>Using a Maxwell's Demon to pump gas</u>

Can a Maxwell's Demon be used to pump gas? Hunt up appropriate references for this discussion.

5. <u>Pumping by momentum transfer from a solid to a gas</u>

If you can cause a solid to vibrate fast enough and with a large enough exertion, can a pumping action be created? How much energy might be required in such a vibrating system? If a solid disk is spun fast enough, can a pumping action be created?

6. <u>Achieving cryopumping temperatures from a subliming solid</u>

If liquid nitrogen, for example, is pumped on by a mechanical pump, the liquid will freeze to a solid. Does the solid expand or contract? If pumping is continued by the mechanical pump, what is the lowest temperature that can be achieved? Consider the effect of a liquid nitrogen jacket around the container holding the subliming solid. Discuss a vacuum dewar between the two chambers.

7. <u>The smallest possible leaks</u>

Distinguish gas leakage from gas permeation. Helium gas permeates glass. Does super-fluid liquid helium leak through glass? Consider a crack or a hole through a solid; is there a lower limit to the magnitude of gas leakage through a hole or a crack in the solid? For a solid wall of vanishingly small thickness, calculate, at least approximately, what this lower limit may be. Now, how about leakage through cracks in thick walls? Consider a crack of the smallest possible dimensions through an infinitely thick wall. In the real world of molecular structure, can you set a limit to the rate of gas flow in steady state? Will the crack plug up? Will the crack heal itself?

BIBLIOGRAPHY

In attempting to compile a concise bibliography, the editors have recognized the impossibility of listing every available source. They have therefore selected a number of references which contain not only pertinent information and ideas but also extensive reading lists of their own.

Texts on Kinetic Theory of Gases

1. E.H. Kennard, Kinetic Theory of Gases, McGraw-Hill, New York, 1938.

2. J.H. Jeans, Kinetic Theory of Gases, Cambridge University Press, London, 1940; Dynamical Theory of Gases, 4th edition, Dover, 1925.

3. S. Chapman and T.G. Cowling, Mathematical Theory of Non-Uniform Gases, Cambridge University Press, London, 1939.

4. L.B. Loeb, Kinetic Theory of Gases, 2nd edition, McGraw-Hill, New York, 1934; 3rd edition, Dover, Peter Smith.

5. R.D. Present, Kinetic Theory of Gases, McGraw-Hill, New York, 1958.

6. S.G. Brush, Kinetic Theory, 2 Vols., Pergamon, London, 1965-1966.

7. T. Wu, Kinetic Equations of Gases and Plasmas, Addison-Wesley, Reading, Mass. 1966.

8. A.R. Hochstim, Kinetic Processes in Gases and Plasmas, Academic, New York, 1968.

9. T.G. Cowling, Molecules in Motion, Harper and Brothers, New York, 1960.

10. F.W. Sears, An Introduction to Thermodynamics, the Kinetic Theory of Gases, and Statistical Mechanics, 2nd edition, Addison-Wesley, Reading, Mass., 1953.

General Reading

11. P.A. Redhead, J.P. Hobson, and E.V. Kornelson, The Physical Basis of Ultrahigh Vacuum, Chapman and Hall, London, 1968.

12. S. Dushman, *Scientific Foundations of Vacuum Technique*, 2nd edition, (J.M. Lafferty, ed.), Wiley, New York, 1962.

13. G. Lewin, *Fundamentals of Vacuum Science and Technology*, McGraw-Hill, New York, 1965.

14. B.D. Power, *High Vacuum Pumping Equipment*, Chapman and Hall, London, 1966.

15. A.E. Barrington, *High Vacuum Engineering*, Prentice-Hall, Englewood Cliffs, N.J., 1964.

16. C.M. Van Atta, *Vacuum Science and Engineering*, McGraw-Hill, New York, 1965.

17. M. Pirani and J. Yarwood, *Principles of Vacuum Engineering*, Reinhold, New York, 1961.

18. A.H. Beck (ed.), *Handbook of Vacuum Physics*, Pergamon, London, 1965-1968.

19. R.W. Roberts and T.A. Vanderslice, *Ultrahigh Vacuum and Its Applications*, Prentice-Hall, Englewood Cliffs, N.J., 1963.

20. J.H. de Boer, *The Dynamical Character of Adsorption*, Oxford University Press, London, 1968.

21. N.W. Robinson, *Physical Principles of Ultrahigh Vacuum Systems and Equipment*, Chapman and Hall, London, 1968.

22. L. Ward and J.P. Bunn, *Introduction to Theory and Practice of High Vacuum Technology*, Butterworth, London, 1967.

Texts with Emphasis on Techniques

23. H.A. Steinherz, *Handbook of High Vacuum Engineering*, Reinhold, New York, 1963.

24. A. Guthrie, *Vacuum Technology*, Wiley, New York, 1963.

25. G.W. Green, *The Design and Construction of Small Vacuum Systems*, Chapman and Hall, London, 1968.

26. W.T.M. Dennis and T.A. Heppell, *Vacuum System Design*, Chapman and Hall, London, 1968.

27. R.H. Berry, P.M. Hall, and M.T. Harris, *Thin Film Technology*, D. Van Nostrand, Princeton, 1968.

28. K.L. Chopra, *Thin Film Phenomena*, McGraw-Hill, New York, 1969.

29. L.I. Maissel and R. Glang (eds.), *Handbook of Thin Film Technology*, McGraw-Hill, New York, 1970.

30. L. Holland (ed.), Thin Film Microelectronics, Chapman and Hall, London, 1965.

31. F. Weber, Elsevier's Dictionary of High Vacuum Science and Technology, Elsevier, Amsterdam, 1968.

32. K. Diels and R. Jaekel, Leybold Vacuum Handbook, (H. Adam and J. Edwards, Trans.), Pergamon Press, London, 1966.

33. O. Heavens, Optical Properties of Thin Solid Films, Academic, New York, 1965.

34. W. S. Spinks, Vacuum Technology, Chapman and Hall, London, 1963.

35. F. Rosebury, Handbook of Electron Tube and Vacuum Techniques, Addison-Wesley, Reading, Mass., 1965.

36. W. Espe, Materials of High Vacuum Technology, Pergamon, London, 1968.

37. A. Roth, Vacuum Sealing Techniques, Pergamon, Longon, 1966.

38. W. F. Brunner, Jr., and T. H. Batzer, Practical Vacuum Techniques, Reinhold, New York, 1965.

39. W. H. Kohl, Materials and Techniques for Electron Tubes, Reinhold, New York, 1960.

40. J. H. Leck, Pressure Measurement in Vacuum Systems, Chapman and Hall, London, 1964.

41. G. Hass and R. E. Thun (eds.), Physics of Thin Films (continuing series), Academic, New York.

42. L. Holland, Vacuum Deposition of Thin Films, Wiley, New York, 1956.

43. L. Holland, The Properties of Glass Surfaces, Wiley, New York, 1964.

44. Vacuum Physics Education Experiments, Edwards High Vacuum International, Crawley, Sussex, England.

45. D. J. Santeler, D. W. Jones, D. H. Holkeboer, and F. Pagano, Vacuum Technology and Space Simulation, NASA SP-105.

46. J. Yarwood, High Vacuum Technique, 4th edition, Chapman and Hall, London, 1967.

47. R. Hawley and A. Maitland, Vacuum as an Insulator, Chapman and Hall, London, 1967.

48. H. Adam (ed.), *Transactions of the Third International Vacuum Congress (Stuttgart, 1965)*, Pergamon, London, 1966, 2 vols.

49. S.P. Wolsky and E.J. Zdanuk (eds.), *Ultra Micro Weight Determination in Controlled Environment*, Wiley, New York, 1969.

50. *Vacuum Microbalance Techniques*, Plenum, 1967, 6 vols.

Journals

51. *Journal of Vacuum Science and Technology* (American Vacuum Society).

52. *Vacuum*, Pergamon Press, Ltd.

53. *Le Vide* (Societe francaise des ingenieurs techniciens du vide).

54. *Vakuum-Technik* (German Vacuum Society).

55. *Surface Science*, North-Holland Publishing Company.

56. *Transactions AVS Vacuum Symposium* (American Vacuum Society, 1959-1964).

57. *Thin Solid Films*, Elsevier Publishing Company.

58. *Thin Films*, Gordon and Breach, Science Publishers, Inc.

APPENDIX

A Limited Glossary of Terms

The following definitions of terms are in conformity with those given in the Glossary of Terms Used in Vacuum Technology, American Vacuum Society, Inc. (Pergamon, New York, 1958), to which the reader is referred for a comprehensive set of terms.

MILLIMETER OF MERCURY: A unit of pressure corresponding to a column of mercury exactly one millimeter high at 0°C under standard acceleration of gravity of 980.665 cm/sec^2. By "mercury at 0°C" is meant a hypothetical fluid having an invariable density exactly 13.5951 g/cm^3. Abbreviated as mmHg.

MICRON OF MERCURY: A unit of pressure equal to 1/1000th of one millimeter of mercury pressure. Abbreviated as μ of Hg or μHg.

TORR: Suggested international standard term to replace the English term millimeter of mercury and its abbreviation mm of Hg (or the French mm de Hg). Both Tor and Torr have been used in Germany, the latter spelling being more common and the one officially adopted by the German Standards Association. The Torr is defined as 1/760 of a standard atmosphere or 1,013,250/760 dynes per square centimeter. This is equivalent to defining the Torr as 1333.22 microbars and differs by only one part in seven million from the International Standard millimeter of mercury. It is recommended that Torr not be abbreviated, however, the abbreciation t has been used.

MILLITORR: A unit of pressure equal to 10^{-3} Torr. Abbreviated as mTorr.

MICROTORR: A unit of pressure equal to 10^{-6} Torr.

MEAN FREE PATH (of any particle): The average distance that a particle travels between successive collisions with the other particles of an ensemble. In vacuum technology the ensemble of particles of interest comprises only the molecules in the gas phase.

ACCOMMODATION COEFFICIENT (for condensation): The ratio of the condensation rate to the impingement rate. The term condensation coefficient is recommended for this ratio.

ACCOMMODATION COEFFICIENT (for free-molecule heat transfer): The ratio of the energy actually transferred between impinging gas molecules and a surface and the energy which would be theoretically transferred if the impinging molecules reached complete thermal equilibrium with the surface.

DIFFUSION COEFFICIENT: The absolute value of the ratio of the molecular flux per unit area to the concentration gradient of a gas diffusing through a gas or a porous medium where the molecular flux is evaluated across a surface perpendicular to the direction of the concentration gradient.

THROUGHPUT: The quantity of gas in pressure-volume units at a specified temperature flowing per unit time across a specified open cross section of a pump or pipe line. The specified temperature may be the actual temperature of the gas or a standard reference temperature. It is recommended that throughput be referred to standard room temperature. The recommended unit of throughput is the Torr · liter per second at 20°C. Other units of throughput in common use are micron liters per second at 25°C and micron cubic feet per minute at 68°F.

    a. Under conditions of steady-state conservative flow, the throughput across the entrance to a pipe is equal to the throughput at the exit. In this case throughput can be defined as the quantity of gas flowing through a pipe in pressure-volume units per unit time at room temperature.

    b. Values of throughput calculated from pressure measurements depend on previous use of calibrated leaks or on theoretical conductance formulas. When untrapped gauges are used to measure pressure, the calculated throughput is influenced by the partial pressure of vapors. However, the effect of backstreaming pump fluid vapor is frequently neglected in calculating the flow of "gas" across the entrance of a vapor pump or through a pipe near the pump. An alternative definition of throughput might therefore state explicitly that pump fluid vapor is not included in the quantity of "gas" involved in the flow calculation.

SPEED: The speed of a pump for a given gas is the ratio of the throughput of that gas to the partial pressure of that gas at a specified point near the mouth (or inlet port) of a pump. Also known as the admittance for a given gas. $S_a = Q_a/P_a$.

CONDUCTANCE:

    a. (Measured value)--The throughput under steady-state conservative conditions divided by the measured difference in pressure between two specified cross sections inside a pumping system. $U = Q/(P_1 - P_2)$

b. (Calculated from long tube formulas)--The throughput under steady-state conservative conditions at a given mean pressure per unit pressure difference across two specified sections in a theoretical pipe of length much greater than the distance between the sections and of uniform or slowly changing cross section as calculated from formulas which neglect entrance and aperture corrections.

c. (Calculated from short tube formulas)--The throughput under steady-state conservative conditions in a pipe line at a given mean pressure per unit pressure difference between two large chambers connected by the pipe line as calculated from formulas which include entrance and aperture corrections.

INDEX OF OPERATIONS, MATERIALS, AND INSTRUMENTS

Numbers are those of the relevant experiment(s) for each entry.

B

Backstreaming, 1.2
Baking, 2.1, 2.2, 3.1, 4.1, 4.2, 4.3, 4.4, 6.1, 6.5
Bourdon gauge, 1.5

C

Calibration, 3.3
Cathode ray tube, 6.4
Charcoal, 4.1, 6.6
Cleaning, 1.4, 2.1, 5.1, 6.2, 6.5
Cold cathode ionization gauge, 1.5, 6.6
Conductance, 1.1, 1.2, 2.2, 2.3, 3.3
Cryosorption pump, 1.3, 1.4, 4.2

D

Diaphragm gauge, 1.5
Diffusion, 2.2, 3.1, 4.3, 4.4
Diffusion pump, 1.2
Dubrovin gauge, 1.5

E

Evaporation, 5.1, 5.2, 6.3

F

Flowmeter, 1.1, 2.3, 3.3
Freeze drying, 6.1
Friction, 6.2

G

Gas flow, 2.2, 2.3, 3.3, 4.4
Gauges
    Bourdon, 1.5
    Calibration, 3.3
    Cold cathode ion, 1.5, 6.6
    Diaphragm, 1.5
    Dubrovin, 1.5
    Linearity, 3.2
    McLeod, 1.1, 1.5, 3.3

INDEX

Mechanical, 1.5
Pirani, 1.5
Pumping action, 3.1, 3.2, 3.3, 4.3
Thermionic ion, 1.5, 3.1, 3.2, 3.3, 4.3, 6.5
Thermocouple, 1.5
U-tube, 1.5
Gettering, 4.3, 6.5, 6.6

I

Ionization gauge, 1.5, 3.1, 3.2, 3.3, 4.3, 6.5
Ionization pumping, 1.4, 3.1, 4.3
Ion Pump, 1.4, 3.1, 4.3
Isotherms, 4.1

L

Lubricants, 6.2

M

Mass spectrometer, 2.2
McLeod gauge, 1.5, 3.3
Mechanical gauge, 1.5
Mechanical pump, 1.1, 1.2, 1.4
Molecular flow, 1.1, 2.3, 3.3
Molecular sieve, 1.3, 4.1, 4.2, 4.3

Monte Carlo technique, 2.3

O

Outgassing, 2.1, 2.2, 3.1, 3.2, 4.1, 4.2, 4.3, 6.2, 6.5

P

Permeation, 2.2, 4.4
Pirani gauge, 1.5
Pumps
    Cryosorption, 1.3, 1.4, 4.2
    Diffusion, 1.2
    Ion, 1.4, 3.1, 4.3
    Mechanical, 1.1, 1.2, 1.4
    Titanium, 4.3
Pump down time, 1.3
Pumping speed, 2.3
    Cryosorption pump, 1.3
    Diffusion pump, 1.2
    Ion gauge, 3.1, 3.3, 4.3
    Ion pump, 1.4
    Mechanical pump, 1.1
    Sorption trap, 4.2
    System, 1.2, 2.2

S

Solder glass, 6.5, 6.6
Sorption, 2.2, 4.1, 4.2
Sorption pumping, 1.3, 1.4, 4.2

INDEX

Sputtering, 5.3, 5.4

Sublimation, 6.1

T

Thermal conductivity gauge, 1.5

Thermionic ionization gauge, 1.5, 3.1, 3.2, 3.3

Thermocouple gauge, 1.5

Thin films, 5.1, 5.2, 5.3, 5.4, 6.3

Titanium, 4.3

Transitional flow, 2.3

Traps, 1.1, 1.5, 2.1, 4.2, 4.3

Triode valve, 6.5

U

Ultimate pressure, 1.1, 1.3, 2.2

U-tube gauge, 1.5

V

Viscous flow, 1.1, 2.3

Other books of interest to you...

Because of your interest in our books, we have included the following catalog of books for your convenience.

Any of these books are available on an approval basis. This section has been reprinted in full from our **material science** catalog.

If you wish to receive a complete catalog of MDI books, journals and encyclopedias, please write to us and we will be happy to send you one.

MARCEL DEKKER, INC.
95 Madison Avenue, New York, N.Y. 10016

material science
including
Polymers, Plastics, Fibers, and Coatings
Metals and Metallurgy
Ceramics and Glass
Vacuum Science

ALTGELT and SEGAL
Gel Permeation Chromatography
edited by KLAUS H. ALTGELT, *Chevron Research Company, Richmond, California*, and LEON SEGAL, *South Regional Research Laboratory, U.S.D.A., New Orleans, Louisiana*
672 pages, illustrated. 1971

Demonstrates the manifold applications of gel permeation chromatography in the field of polymer chemistry. Directed to all research, quality-control, and analytical chemists working with conventional and unconventional polymers and other large molecules in the fields of polymer, cellulose, and petroleum chemistry.

CONTENTS: The sizes of polymer molecules and the GPC separation, F. W. Billmeyer, Jr. and K. H. Altgelt. Gel permeation chromatography column packings – types and uses, D. J. Harmon. Chromatographic instrumentation and detection of gel permeation effluents, E. M. Barrall, II and J. F. Johnson. Peak resolution and separation power in gel permeation chromatography, D. J. Harmon. A review of peak broadening in gel chromatography, R. N. Kelley and F. W. Billmeyer, Jr. Mathematical methods of correcting instrumental spreading in GPC, L. H. Tung. Comparison of different techniques of correcting for band broadening in GPC, J. H. Duerksen. Separation mechanisms in gel permeation chromatography, W. W. Yau, C. P. Malone, and H. L. Suchan. Gel permeation chromatography and thermodynamic equilibrium. E. F. Casassa. Calibration of GPC columns, H. Coll. Data treatment in GPC, L. H. Tung. The overload effect in gel permeation chromatography, J. C. Moore. Gel permeation chromatography using a bio-glas substrate having a broad pore size distribution, A. R. Cooper, J. H. Cain, E. M. Barrall, II, and J. F. Johnson. High resolution gel permeation chromatography – using recycle, K. J. Bombaugh and R. F. Levangie. Gel permeation chromatography with high loads, K. H. Altgelt. Fast gel permeation chromatography, J. N. Little, J. L. Waters, K. J. Bombaugh, and W. J. Pauplis. Extension of GPC techniques, G. Meyerhoff. Phase distribution chromatography (PDC) of polystyrene, R. H. Casper and G. V. Schulz. Apparent and real distribution in GPC (experiments with PMMA samples), K. C. Berger and G. V. Schulz. The instrument spreading correction in GPC. I: The general shape function using a linear calibration curve, T. Provder and E. M. Rosen. The instrument spreading correction in GPC. II: The general shape function using the Fourier transform method with a nonlinear calibration curve, E. M. Rosen and T. Provder. Behavior of micellar solutions in gel permeation chromatography: A theory based on a simple model, H. Coll. Gel permeation analysis of macromolecular association by an equilibrium method, B. F. Cameron, L. Sklar, V. Greenfield, and A. D. Adler. Gel filtration chromatography, B. F. Cameron. Determination of polymer branching with gel permeation chromatography. Abstract of a review, E. E. Drott and R. A. Mendelson. Fractionation of linear polyethylene by gel permeation chromatography. Part III, N. Nakajima. Application of GPC in the study of stereospecific block copolymers, R. D. Mate and M. R. Ambler. Composition of butadiene-styrene copolymers by gel permeation chromatography, H. E. Adams. A direct GPC calibration for low molecular weight polybutadiene, employing dual detectors, J. R. Runyon. Quantitative determination of plasticizers in polymeric mixtures by GPC, D. F. Alliet and J. M. Pacco. Evaluation of pulps, rayon fibers, and cellulose acetate by GPC and other fractionation methods, W. J. Alexander and T. E. Muller. Characterization of the internal pore structures of cotton and chemically modified cottons by gel permeation, L. F. Martin, F. A. Blouin, and S. P. Rowland. Application of GPC to studies of the viscose process. I: Evaluation of the method, L. F. Phifer and J. Dyer. Application of GPC to studies of the viscose process. II: The effects to steeping and alkali-crumb aging, J. Dyer and L. H. Phifer. Gel permeation chromatography calibration. I: Use of calibration curves based on polystyrene in THF and integral distribution curves of elution volume to generate calibration curves for polymers in 2,2,2-trifluoroethanol, T. Provder, J. C. Woodbrey, and J. H. Clark. Modification of a gel permeation chromatograph for automatic sam-

(continued)

material science

ALTGELT and SEGAL (continued)

ple injection and on-line computer data recording, *A. R. Gregges, B. F. Dowden, E. M. Barral, II, and T. T. Horikawa.* Characterization of crude oils by gel permeation chromatography, *H. H. Oelert, D. R. Latham, and W. E. Haines.* Separation and characterization of high-molecular-weight saturate fractions by gel permeation chromatography, *J. H. Weber and H. H. Oelert.* Fractionation of residuals by gel permeation chromatography, *E. W. Albaugh, P. C. Talarico, B. E. Davis, and R. A. Wirkkala.* Combined gel permeation chromatography–NMR techniques in the characterization of petroleum residuals, *F. E. Dickson, R. A. Wirkkala, and B. E. Davis.* A rapid method of identification and assessment of total crude oils and crude oil fractions by gel permeation chromatography, *J. N. Done and W. K. Reid.* Gel permeation analysis of asphaltenes from steam stimulated oil wells, *C. A. Stout and S. W. Nicksic.* GPC separation and integrated structural analysis of petroleum heavy ends, *K. H. Altgelt and E. Hirsch.*

AMERICAN VACUUM SOCIETY
Experimental Vacuum Science and Technology

edited by THE AMERICAN VACUUM SOCIETY EDUCATION COMMITTEE

288 pages, illustrated. 1973

A collection of experiments, which are graded from simple procedures to sophisticated vacuum processes, and designed to aid instructors and introduce students to the basic concepts and techniques of the field of vacuum science. Includes an extensive bibliography to stimulate further investigation. Especially useful for all students and teachers in the many fields of the basic sciences and engineering where vacuum methods and techniques are important.

CONTENTS: **Section 1: Procedures in Vacuum Production and Measurement**, *W. Brunner and H. Patton.* **Section 2: Experiments which Illustrate the Characteristics of the Vacuum Environment:** Demonstration of the outgassing of different vacuum materials, *F. Rosebury.* Comparison of gas evolution phenomena from glass and metal system envelopes during baking, *R. Lawson.* Determination of the net quantity of gas flowing through a cylindrical tube, *K. Busen.* **Section 3: Experiments which Illustrate the Dependence of the Physical Properties of Gases on Gas Density:** Measurement of the pumping action of an ionization gauge, *H. Farber.* Study of the linearity of an ionization gauge, *J. Miller, III.* Calibration of gauges, *C. Morrison.* **Section 4: Experiments which Examine Physical and Chemical Interactions at Surfaces:** Study of the sorption of gases for different gas-sorbent combinations, *K. Wear.* The use of sorbents as traps and pumps, *H. Farber.* Sorption of gases by titanium, *H. Farber.* Investigation of the passage of oxygen across a silver barrier, *K. Busen.* **Section 5: Processes Requiring a Vacuum Environment:** Thin film evaporation, *M. Thomas.* Fabrication of a nichrome resistor, *R. Riegert and G. Breitweiser.* Sputtering, *P. Grosewald.* Ejection patterns in single crystal sputtering, *G. Wehner.* **Section 6: Special Projects:** Study of the sublimation of ice at various pressures, *W. Parker.* Study of friction, *P. McElligott.* Measurement of the mean free path of conduction electrons in silver, *R. Olson and J. Wilson.* Construction and use of a cathode ray tube, *B. Kendall and H. Luther.* Construction of a vacuum triode using solder glass techniques, *J. King and J. Orsula.* Experiments using solder glass techniques, *D. Whitcomb.* **Section 7: Speculations:** Original thought experiments, *M. Carbone.* Provocative ideas and questions, *N. Milleron.*

BEER Liquid Metals: Chemistry and Physics

(Monographs and Textbooks in Material Science Series, Volume 4)

edited by SYLVAN Z. BEER, *Converta Enterprises, Inc., Syracuse, New York*

742 pages, illustrated. 1972

Presents a comprehensive review of the research done on the liquid state of metals, bringing together the latest advances, as well as data previously scattered among a wide variety of publications. Of prime importance to chemists, physicists, research metallurgists, metallurgical engineers, and materials scientists working in the areas of liquid-state theory, the theory of metals, process metallurgy involving liquid metals, and high-temperature chemistry.

CONTENTS: On the thermodynamic formalism of metallic solutions, *C. H. P. Lupis.* Kinetics of evaporation of various elements from liquid iron alloys under vacuum, *R. Ohno.* Relation between thermodynamic and electrical properties of liquid alloys, *D. N. Lee and B. D. Lichter.* The surface tension of liquid metals, *B. C. Allen.* Significant structure theory applied to liquid metals, *S. M. Breitling and H. Eyring.* Diffraction analysis of liquid metals and alloys, *C. N. J. Wagner.* The optical properties of liquid metals, *J. N. Hodgson.* Effect of pressure on the properties of liquid metals, *A. Rapoport.* Sound propagation in liquid metals, *R. T. Beyer and E. M. Ring.* The viscosity of liquid metals, *R. T. Beyer and E. M. Ring.* Magnetic properties of liquid metals, *R. Dupree and E. F. W. Seymour.* Diffusion in liquid metals, *N. H. Nachtrieb.* Electromigration in liquid alloys, *S. G. Epstein.* Electronic nature of liquid metals and liquid metal theory, *J. E. Enderby.* Structure and properties of noncrystalline metallic alloys produced by rapid quenching of liquid alloys, *B. C. Giessen and C. N. J. Wagner.*

material science

BLACK and PRESTON High-Modulus Wholly Aromatic Fibers

(Fiber Science Series, Volume 5)
edited by W. BRUCE BLACK, *Monsanto Textiles Company, Pensacola, Florida* and JACK PRESTON, *Monsanto Textiles Company, Chemstrand Research Center, Durham, North Carolina*
304 pages, illustrated. 1973

Based on a symposium on high-modulus aromatic fibers held by the American Chemical Society in Boston on April 13, 1972. The first formal publication of research which shows the relationship of fiber properties to polymer structure. Extremely significant reading for all fiber scientists and material scientists; plastics scientists and engineers interested in fiber-reinforced plastics; spacecraft and aircraft oriented engineers; and scientists in the industrial fiber, sports equipment, and airframe fields.

CONTENTS: High-modulus wholly aromatic fibers: Introduction to the Symposium and historical perspective, *W. Black*. High-modulus wholly aromatic fibers. I. Wholly ordered polyamide-hydrazines and poly-1,3,4,-oxadiozole-amides, *J. Preston, W. Black and W. Hofferbert, Jr*. High-modulus wholly aromatic fibers. II. Partially ordered polyamide-hydrazides, *J. Preston, W. Black, and W. Hofferbert, Jr*. Self-regulating polycondensations. II. A study of the order present in polyamide-hydrazides derived from terephthaloyl chloride and p-aminobenzhydrazide, *R. Morrison, J. Preston, J. Randall, and W. Black*. Self-regulating polycondensations. III. NMR analysis of oligomers derived from terephthaloyl chloride and p-aminobenzhydrazide, *J. Randall, R. Morrison, and J. Preston*. Some physical and mechanical properties of some high-modulus fibers prepared from all-para aromatic polyamide-hydrazides, *W. Black, J. Preston, H. Morgan, G. Raumann, and M. Lilyquist*. Morphology and crystal structure of wholly aromatic all-para polyamide-hydrazide polymers, *V. Holland*. X-ray study of an all-para wholly aromatic polyamide-hydrazide[a,b], *R. Miller*. Molecular weight characterization of wholly para-oriented, aromatic polyamide-hydrazides and wholly aromatic polyamides, *J. Burke*. Construction and properties of fabrics of high-modulus organic fibers useful for composite reinforcing, *M. Lilyquist, R. DeBrunner, and J. Fincke*. Mechanical properties of a high-modulus polyamide-hydrazide fiber in composites and of the polyamide-hydrazide fiber and fabric composites, *D. Zaukelies and B. Daniels*. Tire cord application of high-modulus fibers derived from polyamide-hydrazides, *G. Raumann and J. Brownlee*. The application of high-modulus fibers to ballistic protection, *R. Laible, F. Figucia, and W. Ferguson*. High-modulus wholly aromatic fibers. III. Random copolymers containing hydrazide and/or carbonamide linkages, *J. Preston, H. Morgan, and W. Black*.

BOLKER Natural and Synthetic Polymers: An Introduction

by HENRY I. BOLKER, *Department of Chemistry, McGill University, Montreal, Quebec*
in preparation. 1973

Presents a unified approach to polymer chemistry, with equal emphasis on natural and synthetic polymers, and is arranged in a logical sequence of topics based on increasing complexity of molecular architecture. Useful as a textbook for a first course in polymer chemistry and as a reference book for workers in the field.

CONTENTS: Introduction • Natural condensation polymers: The linear polysaccharides • Synthetic condensation (step-growth) polymers • Addition (chain-growth) polymers • Stereoregularity in addition polymers • Branched homopolymers: Synthetic and natural • Natural heteropolymers: I. Heteropolysaccharides • Natural heteropolymers: II. Nucleic acids • Copolymers and copolymerization • Cross-linking in synthetic polymers • Natural heteropolymers: III. Polypeptides and proteins • Lignins.

BROWNING Analysis of Paper

by B. L. BROWNING, *The Institute of Paper Chemistry, Appleton, Wisconsin*
352 pages, illustrated. 1969

Provides comprehensive coverage of methods for chemical analysis of paper. Is of value to manufacturers of paper and paper board, suppliers of components or of additives introduced into paper, converters and printers, purchasers and users, librarians, and others concerned with the properties, behavior, and applications of paper that are related to composition.

CONTENTS: Paper as a commodity • Sampling and preparation of sample • Determination of moisture • Fiber analysis • Fiber quality methods • Lignin • Rosin size • Starch • Proteins • Coatings • Waxes and oils • Fillers and white coating pigments • Dyes and colored pigments • Acidity and alkalinity • Residues and impurities • Biological control agents • General identification of additives in paper • Synthetic resins • Wet-strength agents • Polysaccharides and gums • Miscellaneous additives • Noncellulose fibers • Specks and spots • Permanence of paper • Paper in forensic science.

BUTLER, O'DRISCOLL, and SHEN Reviews in Macromolecular Chemistry

(Book Edition)
edited by GEORGE B. BUTLER, *Department of Chemistry, University of Florida,*

(continued)

material science

BUTLER, O'DRISCOLL, and SHEN *(continued)*
Gainesville, and KENNETH F. O'DRISCOLL, *Department of Chemical Engineering, University of Waterloo, Ontario, Canada* and MITCHEL SHEN, *Department of Chemical Engineering, University of California, Berkeley*

Vol. 1 out of print
Vol. 2 388 pages, illustrated. 1968
Vol. 3 430 pages, illustrated. 1969
Vol. 4 428 pages, illustrated. 1970
Vol. 5, Part I see NEUSE and ROSENBERG
Vol. 5, Part II
250 pages, illustrated. 1970
Vol. 6 498 pages, illustrated. 1971
Vol. 7 314 pages, illustrated. 1972
Vol. 8 346 pages, illustrated. 1972
Vol. 9 380 pages, illustrated. 1973

Reviews of the currently published literature for those who wish to keep abreast of the new and rapidly advancing developments in macromolecular chemistry. Of interest to organic and physical chemists, biochemists, engineers, and all students of and research workers in polymer chemistry and related fields.

CONTENTS:

Volume 1: Application of molecular orbital theory to vinyl polymerization, *K. F. O'Driscoll and T. Yonezawa.* Poly(alkylene oxides), *A. E. Gurgiolo.* Polyurethanes, *D. J. Lyman.* Uncatalyzed, uninhibited thermal oxidation of saturated polyolefins, *L. Reich and S. S. Stivala.* Double-strand polymers, *W. De Winter.* Biomedical polymers, *D. J. Lyman.* Gel permeation chromatography with organic solvents, *J. F. Johnson, R. S. Porter, and M. J. R. Cantow.*

Volume 2: Phosphorus-containing polymers: Introduction, *M. Sander and E. Steininger.* Linear polymers with phosphorus in side chains, *M. Sander and E. Steininger.* Linear polymers with phosphorus and carbon in the main chain, *M. Sander and E. Steininger.* Reassessment of the theory of polyesterification with particular reference to alkyd resins, *D. H. Solomon.* Symmetry considerations for stereoregular polymers, *A. M. Liquori.* Copolymerization of vinyl monomers with ring compounds, *R. A. Patsiga.* Application of high-resolution nuclear magnetic resonance to polymer structure determination, I., *K. C. Ramey and W. S. Brey, Jr.* Ten years of polymer single crystals, *D. A. Blackadder.* Thermal degradation of polystyrene, *G. G. Cameron and J. R. MacCallum.*

Volume 3: Phosphorous-containing resins, *M. Sander and E. Steininger.* Inorganic phosphorous polymers, *M. Sander and E. Steininger.* Phosphorylation of polymers, *M. Sander and E. Steininger.* Sulphur-containing polymers, *E. J. Goethals.* Polymer molecular weight distributions, *N. Amundson and D. Luss.* Heteroatom ring-containing polymers, *A. D. Delman.* Molecular theories of rubber-like elasticity and polymer viscoelasticity, *M. Shen, W. F. Hall, and R. E. DeWames.* End-group studies using dye techniques, *S. R. Palit and B. M. Mandal.* Free-radical spin labels for macromolecules, *J. D. Ingham.* The synthesis of thermally stable polymers: A progress report, *J. I. Jones.*

Volume 4: Polymer enzymes and enzyme analogs, *A. S. Lindsey.* Stability of polycarbonate, *A. Davis and J. Golden.* Cross-linking — effect on physical properties of polymers, *L. E. Nielsen.* The synthesis of thermally stable polymeric azomethines by polycondensation reactions, *G. F. D'Alelio and R. K. Schoenig.* On the dehydrochlorination and the stabilization of polyvinyl chloride, *M. Onozuka and M. Asahina.* Recent advances in the development of flame-retardant polymers, *A. D. Delman.* Thermodynamics of polymerization. I, *H. Sawada.*

Vol. 5, Part II: Ring-chain equilibria, *H. Allcock.* Occupied volume of liquids and polymers, *R. Haward.* The application of ESR techniques to high polymer fracture, *H. Kausch-Blecken von Schmeling.* The science of determining copolymerization reactivity ratios, *P. Tidwell and G. Mortimer.* Block polymers and related heterophase elastomers, *G. Estes, S. Cooper, and A. Tobolsky.*

Volume 6: Proton magnetic resonance of molecular interactions in polymer solutions, *K.-J. Liu and J. E. Anderson.* Preparation and polymerization of vinyl heterocyclic compounds, *K. Takemoto.* Catalysis in isocyanate reactions, *K. C. Frisch and L. P. Rumao.* Thermodynamics of polymerization. II. Thermodynamics of ring-opening polymerization, *H. Sawada.* Copolymers of naturally occurring macromolecules, I. *C. Watt.* Molecular configuration and pyrolysis of phenolic-novolaks, *E. L. Winkler and J. A. Parker.* Physical properties of ionic polymers, *E. P. Otocka.* Synthesis and properties of polyphenyls and polyphenylenes, *J. G. Speight, P. Kovacic, and F. W. Koch.* Dependence of flow properties on molecular weight, temperature, and shear, *A. Casale, R. S. Porter, and J. F. Johnson.* Synthesis methods and properties of polyazoles, *V. V. Korshak and M. M. Teplyakov.*

Volume 7: Linear polyquinoxalines, *P. M. Hergenrother.* Nylons—known and unknown. A comprehensive index of linear aliphatic polyamides of regular structure, *H. K. Livingston, M. S. Sioshansi, and M. D. Glick.* Recent advances in polymer fractionation, *L. H. Tung.* Rheology of adhesion, *D. H. Kaelble.* Solvation of synthetic and natural polyelectrolytes, *B. E. Conway.* Hydrogen transfer polymerization with anionic catalysts and the problem of anionic isomerization polymerization, *J. P. Kennedy and T. Otsu.*

Volume 8: Polymerization by carbenoids, carbenes, and nitrenes, *M. Imoto and T. Nakaya.* Collagen and gelatin in the solid state, *I. V. Yannas.* Ring-opening polymerization of cycloolefins, *N. Calderon.* Thermodynamics of polymerization. III, *H. Sawada.* Polymerization of N-vinylcarbazole initiated by metal salts, *M. Biswas and D. Chakravarty.* Vibrational spectroscopy of polymers, *F. J. Boerio and J. L. Koenig.* Polymer compatibility, *S. Krause.*

material science

Volume 9: Mechanical properties of polymers: The influence of molecular weight and molecular weight distribution, *J. Martin, J. Johnson, and A. Cooper*. On the mathematical modeling of polymerization reactors, *W. Ray*. Anionic cyclopolymerization, *C. McCormick and G. Butler*. Carbon-13 NMR of polymers, *V. Mochel*. Thermodynamics of polymerization. IV. Thermodynamics of equilibrium polymerization, *H. Sawada*.

CARROLL   Physical Methods in Macromolecular Chemistry

a series edited by BENJAMIN CARROLL, *Rutgers—The State University, Newark, New Jersey*

Vol. 1   400 pages, illustrated. 1969
Vol. 2   384 pages, illustrated. 1972

A series which reviews why and how analytical methods are used in the study of macromolecules. Each method is critically discussed by experts in the field. Directed to researchers in polymer chemistry, biopolymers, and organic and inorganic chemistry.

CONTENTS:
Volume 1: Surface chemistry and polymers, *M. Rosoff*. Internal reflection spectroscopy, *J. K. Barr and P. A. Flournoy*. Electric properties of synthetic polymers, *E. O. Forster*. Assessing radiation effects in polymers, *P. Y. Feng and E. S. Freeman*. Fluorescence techniques for polymer solutions, *D. J. R. Laurence*. Insoluble polymers: Molecular weights and their distributions, *H. C. Cheung*.
Volume 2: Gel permeation chromatography in polymer chemistry, *D. D. Bly*. Interactions of polymers with small ions and molecules, *D. J. R. Laurence*. Electric properties of biopolymers: Proteins, *E. O. Forster and A. P. Minton*. Thermal methods, *E. P. Manche and B. Carroll*.

CARTER   Essential Fiber Chemistry

(Fiber Science Series, Volume 2)

by MARY E. CARTER, *FMC Corporation, American Viscose Division, Marcus Hook, Pennsylvania*

232 pages, illustrated. 1971

Discusses the chemical and physical structure and properties of ten commercially important fibers. Useful to all chemists interested in the research and development of natural and man-made fibers.

CONTENTS: Cotton • Rayon • Cellulose acetate • Wool • Polyamide • Acrylic fibers • Polyethylene terephthalate • Polyolefins • Spandex • Glass.

CONLEY   Thermal Stability of Polymers

In 2 Volumes

(Monographs in Macromolecular Chemistry Series)

edited by R. T. CONLEY, *Wright State University, Dayton, Ohio*

Vol. 1   656 pages, illustrated. 1970
Vol. 2   in preparation. 1974

CONTENTS: Introduction, *R. T. Conley*. Molecular structure and stability criteria, *R. T. Conley*. The relationship between the kinetics and mechanism of thermal depolymerization, *R. H. Boyd*. Random scission processes, *A. V. Tobolsky, A. M. Kotliar, and T. C. P. Lee*. Fundamental reactions in oxidation chemistry, *P. M. Norling and A. V. Tobolsky*. Thermal and oxidative degradation of polyethylene, polypropylene, and related olefin polymers, *R. H. Hansen*. Thermal and oxidative degradation of natural rubber and allied substances, *E. M. Bevilacqua*. Vinyl and vinylidene polymers, *R. T. Conley and R. Malloy*. Fluorocarbon polymers, *W. W. Wright*. Thermal and thermooxidative degradation of polyamides, polyesters, polyethers, and related polymers, *R. T. Conley and R. A. Gaudiana*. Thermosetting resins, *R. T. Conley*. Thermal and thermooxidative degradation of cellulosic polymers, *R. T. Conley*. Heterocyclic polymers, *G. P. Shulman*. Degradation of inorganic polymers, *J. Economy and J. H. Mason*.

D'ALELIO and PARKER
Ablative Plastics

edited by GAETANO F. D'ALELIO, *Department of Chemistry, University of Notre Dame, Indiana*, and JOHN A. PARKER, *NASA, Ames Research Center, Moffet Field, California*

504 pages, illustrated. 1971

Provides the comprehensive and rational approach required for the design and production of reliable head shields for future space missions. Includes discussions on the various aspects of heat-rejection mode as a function of heating rate; the nature of the heat transfer, both radiative and conductive; and the nature of degrading polymers. A valuable reference for all aerospace scientists, polymer chemists, physicists, and aerodynamic engineers.

CONTENTS: Ablative polymers in aerospace technology, *D. L. Schmidt*. Hypervelocity heat protection—a review of laboratory experiments, *N. S. Vojvodich*. A review of ablative studies of interest to naval applications, *F. J. Koubek*. Structural design and thermal properties of polymers, *G. F. D'Alelio*. Characterization of an epoxy-anhydride ablative system using com-

(continued)

material science

D'ALELIO and PARKER *(continued)*
puter treatment of analytical results, *C. G. Taylor and E. L. Pendleton.* The synthesis and characterization of some potential ablative polymers, *R. Y. Wen, L. F. Sonnabend, and R. Eddy.* Thermal degradation and curing of polyphenylene, *D. N. Vincent.* Thermosetting polyphenylene resin—its synthesis and use in ablative composites, *N. Bilow and L. J. Miller.* Structural ablative plastics, *R. M. Lurie, S. F. D'Urso, and C. K. Mullen.* Prediction of heat shield performance in terms of epoxy resin structure, *G. J. Fleming.* Ablative resins for hyperthermal environments, *B. S. Marks and L. Rubin.* The development of polybenzimidazole composites as ablative heat shields, *R. R. Dickey, J. H. Lundell, and J. A. Parker.* Ablative degradation of a silicon foam, *T. McKeon.* Thermophysical characteristics of high-performance ablative composites, *M. L. Minges.* The design and development of a high-heating rate thermogravimetric analyzer suitable for use with ablative plastics, *A. M. Melnick and E. J. Nolan.* Pyrolysis kinetics of nylon 6-6, phenolic resin and their composites, *H. E. Goldstein.* Pyrolysis-gas chromatography as a tool for studying the degradation of ablative plastics, *R. M. Ross.* Nonequilibrium flow and the kinetics of chemical reactions in the char zone, *G. C. April, R. W. Pike, and E. G. del Valle.* Arc-image testing of ablation materials, *E. M. Liston.* Development and characterization of a radio frequency-transparent ablator, *E. L. Strauss.* Tailoring polymers for entry into the atmosphere of Mars and Venus, *R. G. Nagler.*

DIGGLE   Oxides and Oxide Films in multi-volumes
(The Anodic Behavior of Metals and Semiconductors Series)

edited by JOHN W. DIGGLE, *Research School of Chemistry, The Australian National University, Canberra*

Vol. 1   552 pages, illustrated. 1972
Vol. 2   424 pages, illustrated. 1973

Treats the anodic behavior of metals and semiconductors and peripheral areas in an authoritative and interdisciplinary manner. The initial volumes deal with the physics and chemistry of oxides and oxide films. Of great value for all those involved in electrochemistry, materials science, solid state physics, electrical engineering, metallurgy, corrosion science, semiconductors, and electrochemical technology.

CONTENTS:
Volume 1: Passivation and passivity, *V. Brusić.* Mechanisms of ionic transport through oxide films, *M. J. Dignam.* Electronic current flow through ideal dielectric films, *C. A. Mead.* Electrical double layer at metal oxide-solution interfaces, *S. Ahmed.*

Volume 2: Anodic oxide films: Influence of solid-state properties on electrochemical behavior, *A. Vijh.* Dielectric loss mechanism in amorphous oxide films, *D. M. Smyth.* Porous anodic films in aluminum, *G. C. Wood.* Dissolution of oxide phases, *J. W. Diggle.*

FOURT and HOLLIES
Clothing: Comfort and Function
(Fiber Science Series, Volume 1)

by LYMAN FOURT and NORMAN HOLLIES, *Gillette Research Institute, Rockville, Maryland*

272 pages, illustrated. 1970

A unified review of the present state of knowledge in the science of clothing. Of interest to textile scientists, fiber producers and marketers, textile converters, and garment makers.

CONTENTS: The factors involved in the study of clothing • Clothing considered as a system interacting with the body • Clothing considered as a structural assemblage of materials • Heat and moisture relations in clothing • Physiological and field testing of clothing by wearing it • Physical properties of clothing and clothing materials in relation to comfort • Differences between fibers with respect to comfort • Current trends and new developments in the study of clothing.

FRISCH and REEGEN   Ring-Opening Polymerization
(Kinetics and Mechanisms of Polymerization Series, Volume 2)

edited by KURT C. FRISCH, and SIDNEY L. REEGEN, *Polymer Institute, University of Detroit, Michigan*

544 pages, illustrated. 1969

Covers the polymerization of important classes of cyclic monomers such as ethylene and propylene oxide, alkylenimines, and sulfides, lactones, lactams, cyclic silicone compounds, and cyclic nitrogen containing heterocycles. Of great interest to the industrial, commercial, and academic worlds as it has application to elastomers, coatings, fibers, films, and foams.

CONTENTS: 1,2 Epoxides, *Y. Ishii and S. Sakai.* 1,3 Epoxides and higher epoxides, *P. Dreyfuss and M. P. Dreyfuss.* Cyclic formals, *J. Furukawa and K. Tada.* Cyclic sulfides, *P. Sigwalt.* Alkylenimines, *M. Hauser.* Lactones, *R. D. Lundberg and E. F. Cox.* Lactams, *H. K. Reimschuessel.* Cyclic siloxanes and silazanes, *E. E. Bostick.* Nitrogen-containing heterocyclic compounds, *V. Kargin and V. Kabanov.* N-carboxy-α-amino acid anhydrides, *Y. Shalitin.*

FRISCH and SAUNDERS
Plastic Foams
(Monographs on Plastic Series, Volume 1)

edited by KURT C. FRISCH, *University of Detroit, Michigan,* and JAMES H. SAUNDERS, *Monsanto Company, Pensacola, Florida*

Part I   464 pages, illustrated. 1972
Part II 704 pages, illustrated. 1973

Gives an integrated picture of the fundamental principles, technology, and applications of foams, and offers a thorough treatment of specific types of plastic foams. Emphasis is placed on the newer trends in this science.

Of particular value to chemists and engineers engaged in research and development, and marketing and production personnel in the polymer and plastics industry.

CONTENTS:
Part I: Introduction, *K. C. Frisch.* The mechanism of foam formation, *J. H. Saunders and R. H. Hansen.* Flexible polyurethane foams, *G. T. Gmitter, H. J. Fabris, and E. M. Maxey.* Sponge rubber and latex foam, *R. L. Zimmerman and H. R. Bailey.* Polyolefin foams, *D. J. Sundquist.* Polyvinyl chloride foams, *A. C. Werner.* Silicone foams, *H. L. Vincent.* Testing of cellular materials, *R. A. Stengard.*
Part II: Rigid urethane foams, *J. K. Backus and P. G. Gemeinhardt.* Polystyrene and related thermoplastic foams, *A. R. Ingram and J. Fogel.* Phenolic foams, *A. J. Papa and W. R. Proops.* Urea-formaldehyde foams, *K. C. Frisch.* Epoxy-resin foams, *H. Lee and K. Neville.* New high-temperature-resistant plastic foams, *E. E. Hardy and J. H. Saunders.* Miscellaneous foams, *K. C. Frisch.* Inorganic foams, *M. Wismer.* Effects of cell geometry on foam performance, *R. H. Harding.* Thermal decomposition and flammability of foams, *P. E. Burgess, Jr. and C. J. Hilado.* Foams in transportation, *M. Kaplan and L. M. Zwolinski.* Architectural uses of foam plastics, *S. C. A. Paraskevopoulos.* Military and space applications of cellular materials, *R. J. F. Palchak.*

GARG, SVALBONAS, and GURTMAN
Analysis of Structural Composite Materials
(Monographs and Textbooks in Material Science Series, Volume 6)

by SABODH GARG, *Systems, Science and Software Company, La Jolla, California,* VYTAS SVALBONAS, *Grumman Aerospace Corporation, Bethpage, New York,* and GERRY GURTMAN, *Systems, Science and Software Company, La Jolla, California*

552 pages, illustrated. 1973

Compares various theories for the static and dynamic analysis of structural composite materials. Deals with the elastic properties of laminated composites and particulate and unidirectional fiber reinforced composites, composite strength, and stress wave propagation. May be used as a textbook for a graduate course in composites and is of interest to researchers and analysts in any industry that uses composites.

CONTENTS: Why composites? • Simple analytic models • Elasticity analyses • Bounds on elastic properties by energy methods • Multilayer laminates • Non-statistical models of composite strength • Statistical tensile strength of fiber and fiber bundles • Composite tensile-strength models • Cumulative weakening model including stress concentrations • Compressive strength of composites • Theory of breaking kinetics • Introduction to elastic wave propagation in composites • Approximate analysis techniques for stress wave propagation in composites • Application of continuum mixture theories to the study of elastic wave propagation in composite materials • Shock waves in composite materials.

HAM Vinyl Polymerization
In 2 Parts
(Kinetics and Mechanisms of Polymerization Series, Volume 1)

edited by GEORGE E. HAM, *Geigy Chemical Corporation, Ardsley, New York*

Part I   560 pages, illustrated. 1967
Part II 432 pages, illustrated. 1969

"The book is a good introduction to the series. It has provided a sound basis for subsequent volumes and should serve as an important reference text for students and researchers in polymer chemistry."—B. D. Gesner, Bell Telephone Labs., *SPE Journal*

"The book is highly recommended."—Arthur Tobolsky, *The American Scientist*

CONTENTS:
Part I: General aspects of free-radical polymerization, *G. E. Ham.* The mechanism of cyclopolymerization of nonconjugated diolefins, *W. E. Gibbs and J. M. Barton.* Styrene, *M. H. George.* Mechanism of vinyl acetate polymerization, *M. K. Lindemann.* Polymerization of vinyl chloride and vinylidene chloride, *G. Talamini and E. Peggion.* Occlusion phenomena in the polymerization of acrylonitrile and other monomers, *A. D. Jenkins.* Polymerization of acrolein, *R. C. Schulz.* Heats of polymerization and their structural and mechanistic implications, *R. M. Joshi and B. J. Zwolinski.*
Part II: Mechanism of emulsion polymerization, *J. W. Vanderhoff.* Elucidation of emulsion polymerization mechanism based upon copolymer studies, *W. F. Fowler, Jr.* Mechanism of the emulsion polymerization of ethylene, *H. K. Stryker, G. J. Mantell, and A. F. Helin.* Mechanism of stereospecific polymerization of propylene, *W. "E" Smith.* Anionic polymerization,

(continued)

material science

HAM (continued)
M. Morton. Mechanisms of cationic polymerization, Z. Zlámal. Radiation-induced polymerization, Y. Tabata.

HENCH and DOVE Physics of Electronic Ceramics

In 2 Parts

(Ceramics and Glass: Science and Technology Series, Volume 2)

edited by LARRY L. HENCH and DEREK B. DOVE, *College of Engineering, University of Florida, Gainesville.*

Part A 584 pages, illustrated. 1971
Part B 576 pages, illustrated. 1972

A highly useful treatise which deals with the physical basis for the behavior of electronic ceramics.

Fundamental physical theories describing each type of electronic ceramics are presented, with discussions included that relate the theories to the applications of the materials. Of special value to graduate students who have had a course in modern physics and also of interest to materials scientists and engineers in electronics, communications, and ceramics.

CONTENTS:

Part A: Quantum mechanics and ceramics, *J. C. Slater*. Band structure and electronic properties of ceramic crystals, *D. Adler*. Electrical conduction in low mobility materials, *I. Bransky and N. M. Tallan*. Defect structure and electronic properties of ceramics, *R. W. Vest*. Conduction domains in solid mixed conductors and electrolytic domain of calcia stabilized zirconia, *J. Patterson*. Semiconducting glasses, *J. D. Mackenzie*. Electronic processes in amorphous semiconductors, *E. A. Davis*. Heterogeneous semiconducting glasses, *H. F. Schaake*. The determination of local order in amorphous semiconducting films, *D. B. Dove*. Negative capacitance effects in amorphous semiconductors, *M. Allen, P. Walsh, and W. Doremus*. Some conduction phenomena in amorphous materials, *K. L. Chopra*. Applications of thin film dielectrics in microelectronics, *N. N. Axelrod*. Substructure and electrical conduction in amorphous thin films, *N. Fuschillo and A. D. McMaster*. Structure of surface defects, *D. L. Stoltz and J. J. Hren*. Electronic surface properties, *P. Mark*. Electron spin resonance and defects in solids, *W. S. Brey, R. B. Gammage, and Y. P. Virmani*. Theory of linear dielectrics, *A. D. Franklin*. Polycrystalline insulators, *H. C. Graham and N. M. Tallan*. Electrical conduction in glass and glass-ceramics, *D. L. Kinser*. Dielectric breakdown of ceramics, *G. C. Walther and L. L. Hench*.

Part B: Some structural mechanisms in ferroelectricity, *R. Pepinsky*. Thermodynamic phenomenology of ferroelectricity in single crystal and ceramic systems, *L. E. Cross*. Dynamical effects in solid state phase transformations, *J. D. Axe*. Theory of antiferromagnetism and ferrimagnetism, *J. B. Goodenough*. Microstructure and processing of ferrites, *F. J. Schnettler*. Microwave garnet compounds, *G. R. Harrison and L. R. Hodges, Jr*. The optical absorption of glasses, *N. J. Kreidl*. Light scattering from glass, *J. J. Hammel*. Electro-optical and magneto-optical effects, *Y. R. Shen*. The influence of the composition of the gain of Nd-doped glasses, *C. F. Rapp*. Solid state reactions in the preparation of zircon stains, *R. A. Eppler*. Computer color control for ceramic wall tile, *W. K. Culbreth, Jr*. Ceramics and glasses – some uses in the communications industry, *D. G. Thomas*.

HENCH and GOULD Characterization of Ceramics

(Ceramics and Glass: Science and Technology Series, Volume 3)

edited by LARRY L. HENCH and ROBERT W. GOULD, *University of Florida, Gainesville*

672 pages, illustrated. 1971

Focuses on the two major directions which comprise the distinct discipline of ceramic characterization: the exploration of the factors that control the properties of the final product and the rapid development of high resolution analytical techniues used for ceramic materials. A particularly timely textbook for an advanced undergraduate or graduate materials science curriculum. Valuable for all materials scientists and ceramic and materials engineers working on the development of an effective and economical materials characterization program.

CONTENTS: Introduction to the characterization of ceramics, *L. L. Hench*. **Part 1: Chemical Analysis:** General analytical chemistry, *P. Rankin*. X-ray spectroscopy, *R. W. Gould*. Atomic absorption flame spectrometry, *J. D. Winefordner*. **Part 2: Phase State and Structure:** X-ray diffraction, *R. W. Gould*. Transmission electron microscopy and electron diffraction, *C. F. Tufts*. Analysis of microstructural defects, *R. W. Newman*. Petrographic analysis, *V. D. Fréchette*. Thermal analysis, *R. K. Ware*. Point defect analysis, *W. J. James and G. Lewis*. **Part 3: Size, Shape, Strain, and Surface of Powders:** Physical characterization, *D. R. Lankard and D. E. Niesz*. Small angle x-ray scattering, *R. W. Gould*. X-ray line profile analysis, *R. W. Gould*. Scanning electron microscopy, *S. R. Bates*. Light scattering, *J. H. Boughton*. Characterization of powder surfaces, *L. L. Hench*. **Part 4: Microstructure:** Electron microprobe, *G. Lewis*. Quantitative stereology, *R. T. DeHoff*. Applied stereology, *S. W. Freiman*. **Part 5: Surfaces:** Characterization of ceramic surfaces, *L. Berrin and R. C. Sundahl*.

material science

KATON Organic Semiconducting Polymers

(Monographs in Macromolecular Chemistry Series)

edited by J. E. KATON, *Miami University, Oxford, Ohio*

328 pages, illustrated. 1968

CONTENTS: Basic physics of semiconductors, *D. E. Hill.* Theoretical aspects of the electronic behavior of organic macromolecular solids, *H. A. Pohl.* Recent experimental aspects of the electronic behavior of organic macromolecular solids, *S. Kanda and H. A. Pohl.* Semiconducting organic polymers containing metal groups, *B. A. Bolto.* Semiconducting biological polymers, *D. D. Eley.*

KETLEY The Stereochemistry of Macromolecules

In 3 Volumes

edited by A. D. KETLEY, *W. R. Grace & Co., Clarksville, Maryland*

Vol. 1 424 pages, illustrated. 1967
Vol. 2 400 pages, illustrated. 1967
Vol. 3 476 pages, illustrated. 1968

CONTENTS:

Volume 1: Ziegler-Natta polymerization: Catalysts, monomers, and polymerization procedures, *D. O. Jordan.* The mechanism of Ziegler-Natta catalysis. I. Experimental foundations, *D. F. Hoeg.* Mechanism of Ziegler-Natta polymerization. II. Quantum-chemical and crystal-chemical aspects, *P. Cossee.* Copolymerization of olefins by Ziegler-Natta catalysts, *I. Pasquon, A. Valvassori, and G. Sartori.* Polymerization of dienes by Ziegler-Natta catalysts, *W. Marconi.* Manufacture and commercial applications of stereoregular polymers, *M. Compostella.*

Volume 2: Stereospecific polymerization of vinyl-type monomers and dienes by alkali-metal-based catalysts, *D. Braun.* Stereospecific polymerization of vinyl ethers, *A. D. Ketley.* Ionic polymerization of aldehydes, ketones, and ketenes, *G. F. Pregaglia and M. Binaghi.* Stereospecific polymerization of epoxides, *T. Tsuruta.* Stereochemistry of free-radical polymerizations, *W. Cooper.* Conformational effects induced in polymers by rigid matrices, *N. Marans.* Simple stereoregular polymers in biological systems, *J. N. Baptist.*

Volume 3: Chain conformation and crystallinity, *P. Corradini.* High-resolution nuclear magnetic resonance of synthetic polymers, *J. C. Woodbrey.* Vibrational analyses of the infrared spectra of stereoregular polypropylene, *T. Miyazawa.* Optically active stereoregular polymers, *M. Farina and G. Bressan.* Physical properties of stereoregular polymers in solid state, *J. F. Johnson and R. S. Porter.* Properties of synthetic linear stereoregular polymers in solution, *V. Crescenzi.* Macromolecules as information storage systems, *A. M. Liquori.* Automata theories of hereditary tactic copolymerization, *H. H. Pattee.* Effect of microtacticity on reactions of polymers, *M. M. van Beylen.* Degradation of stereoregular polymers, *H. H. G. Jellinek.*

KURYLA and PAPA Flame Retardancy of Polymeric Materials

a series edited by WILLIAM C. KURYLA and ANTHONY J. PAPA, *Union Carbide Corporation, South Charleston, West Virginia*

Vol. 1 352 pages, illustrated. 1973
Vol. 2 296 pages, illustrated. 1973

A series concerned with the various modes of rendering polymeric materials fire resistant, which emphasizes specific reagents and techniques in use today. Treats each class of polymer separately to aid the fabricator in gaining an understanding of the specific problems associated with its flammability characteristics, and to review the science and practical solution to its flame retardancy. Of special benefit to industrial polymer chemists, plastics engineers, and fabricators of polymeric materials.

CONTENTS:

Volume 1: Available flame retardants, *W. Kuryla.* Inorganic flame retardants and their mode of action, *J. Pitts.* Fire retardation of polyvinyl chloride and related polymers, *M. O'Mara, W. Ward, D. Knechtges, and R. Meyer.* Fire retardation of wool, nylon, and other natural and synthetic polyamides, *G. Crawshaw, A. Delman, and P. Mehta.*

Volume 2: Fire retardation of polystyrene and related thermoplastics, *R. Lindemann.* Fire retardation of polyethylene and polypropylene, *R. Schwarz.* Flame retardation of natural and synthetic rubbers, *H. Fabris and J. Sommer.* Flame retardancy of phenolic resins and urea- and melamine-formaldehyde resins, *N. Sunshine.*

LEFEVER Aspects of Crystal Growth

(Preparation and Properties of Solid State Materials Series, Volume 1)

edited by R. A. LEFEVER, *Sandia Laboratories, Albuquerque, New Mexico*

Vol. 1 296 pages, illustrated. 1971

Concerns certain aspects of the growth and properties of single crystals. Directed to both beginning and experienced crystal growers, material scientists, and solid state physicists.

CONTENTS: A review of the preparation of single crystals by fused melt electrolysis and some general properties, *W. Kunnmann.* The role of mass transfer in crystallization processes, *W. Wilcox.* Exploratory flux crystal growth, *A. Chase.*

material science

LOWRY   Markov Chains and Monte Carlo Calculations in Polymer Sciences
(Monographs in Macromolecular Chemistry Series)
edited by GEORGE G. LOWRY, *Western Michigan University, Kalamazoo*
344 pages, illustrated. 1970

Written for the polymer chemist who, although not primarily concerned with mathematical theories, desires a working knowledge of the topics treated. Begins with an introduction to the principles involved and later exemplifies some significant applications of Markov chain theory and Monte Carlo methods.

CONTENTS: Introduction: Deterministic and stochastic approaches, *G. G. Lowry*. Markov chains, *J. Myhre*. Monte Carlo methods, *M. Fluendy*. Polymer conformation as a Markov chain problem, *J. Kinsinger*. Polymer conformation and the excluded-volume problem, *S. Windwer*. Higher order Markov chains and statistical thermodynamics of linear polymers, *J. Mazur*. Copolymer composition and tacticity, *F. P. Price*. Molecular-weight distributions, *G. G. Lowry*.

McCULLOUGH   Concepts of Fiber-Resin Composites
(Monographs and Textbooks in Material Science Series, Volume 2)
by R. L. MCCULLOUGH, *Boeing Scientific Research Laboratories, Seattle, Washington*
128 pages, illustrated. 1971

Presents basic concepts of composite material systems. Introduces the study of composite systems by discussing where and how composite materials are used and how their components are selected. A valuable reference for materials scientists, and students and research management engaged in the exploration of composite materials.

CONTENTS: Materials • Composite structures • Composite properties • The interphase region • Synopsis.

MAY and TANAKA   Epoxy Resins: Chemistry and Technology
edited by CLAYTON MAY, *Shell Development Company, Emeryville, California* and YOSHIO TANAKA, *Research Institute for Polymers and Textiles, Yokohama, Japan*
704 pages, illustrated. 1973

Brings together the contributions of a number of outstanding researchers in the field of epoxy resins. Not only emphasizes the chemistry and technology of epoxy resins, but also deals with many industrial applications. Of great value for polymer chemists and technicians, and a wide variety of engineers.

CONTENTS: Introduction to epoxy resins, *C. May*. Synthesis and characteristics of epoxides, *Y. Tanaka, A. Okada, and I. Tomizuka*. Epoxide-curing reactions, *Y. Tanaka and T. Mika*. Curing agents and modifiers, *T. Mika*. Properties of cured resins, *D. Kaelble*. Epoxy-resin adhesives, *A. Lewis and R. Saxon*. Epoxy-resin coatings, *G. Somerville and I. Smith*. Electrical and electronic applications, *A. Breslau*. Epoxy laminates, *J. DelMonte*. Polymer stabilizers and plasticizers, *W. Port*. Analysis of epoxides and epoxy resins, *H. Jahn and P. Goetzky*. Toxicity, hazards, and safe handling, *H. Borgstedt and C. Hine*.

MILLICH and CARREHER   Interfacial Synthesis
edited by FRANK MILLICH, *Department of Chemistry, University of Missouri, Kansas City*, and CHARLES E. CARREHER, JR., *Department of Chemistry, University of South Dakota, Vermillion*
in preparation. 1973

Summarizes the accomplishments in interfacial synthesis to date. Speculates on mechanism, discusses the complex matter of synthetic control, and points out the beneficial aspects of interfacial synthesis in comparison to alternative methods of synthesis. Of fundamental concern to graduate students and teachers of organic chemistry, physical chemists, macromolecular biochemists, and polymer chemists who engage in interfacial synthesis.

CONTENTS: Stirring in organic chemical synthesis, *J. Rushton*. High-speed stirring in interfacial synthesis, *J. Rushton*. Problems and solutions in kinetics and mechanisms, *J. Bradbury and P. Crawford*. Interface effects on chemical reaction rate, *F. MacRitchie*. Liquid-vapor interfacial polycondensations, *L. Sokolov*. Copolycondensation and macroscopic kinetics, *L. Sokolov and V. Nikonov*. The role of the particle-water interface in polymerization, *J. Gardon*. Biochemical reactions at an interface, *R. Baier and D. Cadenhead*. Commercial application of interfacial synthesis, *E. Oliver and Y. Yen*. Polycarboxylic esters, *S. Temin*. Polycarbonates, *H. Vernaleken*. Polycondensations with carbon suboxide, *I. Daniewska*. Polyamides, *V. Nikonov and V. Savinov*. Polyesteramides, *I. Panayotov*. Polyurethanes, *T. Tanaka and T. Yokoyama*. Polyureas, *K. Stueben and A. Barnabeo*. Polyphosphonates, polyphosphates, and polyphosphites, *F. Millich, J. Teague, K. Lambing, and D. Hackathorn*. Other phosphorus containing polymers, *C. Carreher, Jr*. Organometal-

material science

lic polymers, C. *Carreher, Jr.* Modification of natural polymers by interfacial methods, *M. Horio.* Interfacial modifications of poly(vinyl alcohol) and related polymers, *M. Tsuda.* High temperature resistant polymers made by interfacial polymerization, *H. Mark and S. Atlas.*

MYERS and LONG Characterization of Coatings: *Physical Techniques*

In 2 Parts

(Treatise on Coatings Series, Volume 2)

edited by RAYMOND R. MYERS, *Paint Research Institute, Kent State University, Ohio,* and J. S. LONG, *Department of Chemistry, University of Southern Mississippi, Hattiesburg*

Part I 696 pages, illustrated. 1969
Part II in preparation. 1973

Explores the scientific frontier that has developed since the appearance of Mattiello's treatise on coatings. Emphasizes the urgent need of the paint industry to master new technological concepts and instrumental techniques to match the rapid pace of development of its products. Written for the working paint scientist, the laboratory assistant, technician, and superintendent. Also a valuable reference for the formulator and personnel engaged in the production of raw materials for the paint industry.

CONTENTS:

Part I: The intrinsic properties of polymers, *A. Tawn.* Surface areas, *D. Gans.* Adhesion of coatings, *A. Lewis and L. Forrestal.* Mechanical properties of coatings, *P. Pierce.* The ultimate tensile properties of paint films, *R. Evans.* Gas chromatography, *J. Haken.* Thermoanalytical techniques, *P. Garn.* Microscopy in coatings and coating ingredients, *W. Lind.* Radioactive isotopes, *G. Coe.* Infrared spectroscopy *C. Smith.* Ultraviolet and visible spectroscopy, *F. Spagnolo and E. Scheffer.* Color of polymers and pigmented systems, *G. Ingle.* Photoelastic coatings, *A. Blumstein.*

Part II: Dielectric properties, *S. Negami.* Gel permeation chromatography, *K. Boni.* Infrared Fourier transform spectroscopy, *M. Low.* Interfacial energetics, *D. Gans.* Nuclear magnetic resonance, *M. Levy.* Particle sizing, *B. DeWitt.* Scanning electron microscopy, *L. Princen.* Solubility, *J. Gordon and J. Teas.* Transport properties, *G. Park.* Viscometry, *K. Oesterle.* X-ray analysis, diffraction, and emission, *R. Scott.*

MYERS and LONG Film-Forming Compositions

In 3 Parts

(Treatise on Coatings Series, Volume 1)

edited by RAYMOND R. MYERS, *Paint Research Institute, Kent State University, Ohio,* and J. S. LONG, *Department of Chemistry, University of Southern Mississippi, Hattiesburg*

Part I 584 pages, illustrated. 1967
Part II 448 pages, illustrated. 1968
Part III 608 pages, illustrated. 1972

Devoted to materials which form, or aid the formation of, continuous films. Discusses vehicles and resins, placing considerable emphasis on procedures for developing a suitable vehicle for conveying dissolved or suspended solids to a substrate and imparting to the surface those protective and decorative properties for which the coating was designed. Of inestimable value to chemists, formulators, laboratory assistants, technicians, and production superintendents of the coatings industry. Also valuable for laboratory personnel of raw material suppliers, chemists in the plastics and similar industries, and as an excellent reference treatise for libraries.

Part I: Acrylic ester emulsions and water-soluble resins, *G. Allyn.* Acrylic ester resins, *G. Allyn.* Alkyd resins, *W. M. Kraft, E. G. Janusz, and D. Sughrue.* Asphalt and asphalt coatings, *S. H. Alexander.* Chlorinated rubber, *H. E. Parker.* Driers, *W. J. Stewart.* Epoxy resin coatings, *G. R. Somerville.* Hydrocarbon resins and polymers, *D. F. Koenecke.* Hydrocarbon solvents, *W. W. Reynolds.* Natural resins, *C. L. Mantell.* Polyethers and polyesters, *A. C. Filson.* Urethane coatings, *A. Damusis and K. C. Frisch.* Vehicle manufacturing equipment, *A. F. Steioff.*

Part II: Styrene-butadiene latexes in protective and decorative coatings, *F. A. Miller.* Starch polymers and their use in paper coating, *T. F. Protzman and R. M. Powers.* Cellulose esters and ethers, *J. B. G. Lewin.* Drying oils—modifications and use, *A. E. Rheineck and R. O. Austin.* Paint and painting in art, *S. Rees Jones.* Rosin and modified rosins and resins, *C. L. Mantell.* Urea and melamine resins, *H. P. Wohnsiedler and W. L. Hensley.* Vinyl resins *W. H. McKnight and G. S. Peacock.* Vinyl emulsions, *H. D. Cogan and A. L. Mantz.*

Part III: Dimer acids in surface coatings, *J. Boylan.* Emulsion technology, *L. Princen.* Phenolic resins for coatings, *S. Richardson and W. Wertz.* Plasticization of coatings, *F. Ball.* Fatty polyamides and their applications in protective coatings, *D. Wheller and D. Peerman.* Polycarbonate resins, *D. Fox and K. Goldblum.* Reactive polyesters, *F. Ball.* Varnishes, *L. Montague.* Reactive silanes as adhesion promoters to hydrophilic surfaces, *E. Plueddemann.* Surface-active agents, *T. Ginsberg.* Shellac, *J. Martin.* Tall oil in surface coatings, *R. Perez.* Silicones in protective coatings, *L. Brown.*

material science

MYERS and LONG Formulations
(Treatise on Coatings Series, Volume 4)
edited by RAYMOND R. MYERS, Department of Chemistry, Kent State University, Ohio, and J. S. LONG, University of Southern Mississippi, Hattiesburg
Part I in preparation. 1973

NEUSE and ROSENBERG
Metallocene Polymers
(Reviews in Macromolecular Chemistry Series, Volume 5, Part I)
by EBERHARD NEUSE, F. J. Weck Company, City of Industry, California, and HAROLD ROSENBERG, Air Force Materials Laboratory, Wright-Patterson Air Force Base, Ohio
170 pages, illustrated. 1970

Presents a comprehensive and critical account of the progress made in the synthesis and characterization of metallocene polymers.

CONTENTS: Introduction • Macromolecular compounds with pendent metallocenyl groups • Macromolecular compounds with intrachain metallocenylene groups • Conclusions.

O'CONNOR Instrumental Analysis of Cotton Cellulose and Modified Cotton Cellulose
(Fiber Science Series, Volume 3)
edited by ROBERT T. O'CONNOR, Agricultural Research Service, U.S.D.A., New Orleans, Louisiana
512 pages, illustrated. 1972

Describes the applications of instrumental procedures specifically developed by the textile chemist to meet today's demands. Particularly geared to textile chemists and others in the textile industry, and to paper and wood manufacturers. Also of interest to polymer chemists, analytical chemists, and other researchers involved with the processes by which fibers are blended and modified.

CONTENTS: Elemental analysis: Detection, identification, and quantitative determination of metals and nonmetallic elements, R. T. O'Connor. Infrared spectroscopy and physical properties of cellulose, C. Y. Liang. Light microscopy in the study of cellulose, M. L. Rollins and I. V. de Gruy. Electron microscopy of cellulose and cellulose derivatives, M. L. Rollins, A. M. Cannizzaro, and W. R. Goynes. Instrumental methods in the study of oxidation, degradation, and pyrolysis of cellulose, P. K. Chatterjee and R. F. Schwenker, Jr. X-ray diffraction, V. W. Tripp and C. M. Conrad. Wide-line nuclear magnetic resonance spectroscopy, R. A. Pittman and V. W. Tripp. The infrared spectra of chemically modified cotton cellulose, R. T. O'Connor.

PEARL The Chemistry of Lignin
by IRWIN A. PEARL, The Institute of Paper Chemistry, Appleton, Wisconsin
360 pages, illustrated. 1967

CONTENTS: Nebulous concept of lignin • Isolation of lignin • Chemical structure of lignin • Biosynthesis and formation of lignin • Reactions of lignin in major pulping and bleaching processes • Chemical reactions of lignin • Physical properties of lignin and its preparations • Biological decomposition of lignin • Thermal decomposition of lignin • Linkage of lignin in the plant • Utilization of lignin and its preparations.

PETERLIN Plastic Deformation of Polymers
edited by A. PETERLIN, Research Triangle Institute, Research Triangle Park, North Carolina
318 pages, illustrated. 1971

Explores the actual mechanism of deformation that occurs during the formation of fibers and films. Focuses primarily on three topics: what happens on a molecular, crystalline, and supercrystalline level; how the original structure influences the deformation; and what determines the useful mechanical properties of fibers and films. A valuable tool for the staff of research and development laboratories working with plastics, rubbers, fibers, and films, and for students and faculty involved in material science and polymer chemistry.

CONTENTS: Infrared studies of the role of monoclinic structure in the deformation of polyethylene, Y. Kikuchi and S. Krimm. Infrared studies of drawn polyethylene. Part I. Changes in orientation and conformation of highly drawn linear polyethylene, W. Glenz and A. Peterlin. Structure of oriented polyacrylonitrile films, J. L. Koenig, L. E. Wolfram, and J. G. Grasselli. Morphology and deformation behavior of "row-nucleated" polyoxymethylene film, C. A. Garber and E. S. Clark. Plastic deformation of polypropylene. VI. Mechanism and properties, F. J. Baltá-Calleja and A. Peterlin. Polyethylene crystallized under the orientation and pressure of a pressure capillary viscometer. Part I., J. H. Southern and R. S. Porter. Heat relaxation of drawn polyoxymethylene, A. Siegmann and P. H. Geil.

material science

Retraction of cold-drawn polyethylene and polypropylene, *D. Hansen, W. F. Kracke, and J. R. Falender*. Electron spin resonance studies of free radicals in mechanically loaded nylon 66, *G. S. P. Verma and A. Peterlin*. Transition from linear to nonlinear viscoelastic behavior. Part I. Creep of polycarbonate, *I. V. Yannas and A. C. Lunn*. Yielding behavior of glassy polymers. III. Relative influences of free volume and kinetic energy, *K. C. Rusch and R. H. Beck, Jr*. Yielding of quenched and annealed polymethyl methacrylate, *D. H. Ender*. Yield phenomenon in oriented polyethylene terephthalate, *M. Parrish and N. Brown*. Electron paramagnetic resonance investigation of molecular bond rupture due to ozone in deformed rubber, *K. I. DeVries, E. R. Simonson, and M. L. Williams*. Factors affecting the depth of draw in a cold-forming operation, *H. L. Li, P. J. Koch, D. C. Prevorsek, and H. J. Oswald*. Quantitative structural characterization of the mechanical properties of isotactic polypropylene, *R. J. Samuels*.

RAVVE   Organic Chemistry of Macromolecules: An Introductory Textbook

by A. RAVVE, *Continental Can Company, Chicago, Illinois*

512 pages, illustrated. 1967

CONTENTS: **Part I: Introduction:** Historical introduction and definitions • Physical properties of macromolecules • Molecular weights of polymers • **Part II: Polymerization Reactions-Mechanisms:** Addition polymerization: Mechanism of free-radical polymerization • Ionic polymerization • Polymerization with the aid of complex catalysts • Stereospecific polymerization • Bulk, solution, suspension, and emulsion polymerization. **Part III: Common Addition Polymers:** Macroalkanes • Polymers and copolymers from dienes and polyenes • Styrene and styrene-like polymers and polyacrylics • Halogen-bearing addition polymers and vinyl esters and ethers. **Part IV: Condensation Polymers:** Mechanism of polycondensation reactions • polyesters • Polyamides • Polycarbamates, polyureas, and polycarbodiimides • Phenoplasts • Aminoplasts • Ladder and semiladder polymers. **Part V: Naturally Occurring Polymers:** Polysaccharides • Proteins • Polynucleotides. **Part VI: Reactions of Polymers:** Graft and block copolymers • Reactions of polymers • Degradation of polymers.

REICH and STIVALA   Autoxidation of Hydrocarbons and Polyolefins: Kinetics and Mechanisms

by LEO REICH, *Picatinny Arsenal, Dover, New Jersey*, and SALVATORE S. STIVALA, *Department of Chemistry, Stevens Institute of Technology, Hoboken, New Jersey*

544 pages, illustrated. 1969

CONTENTS: Introduction • Oxidation of simple hydrocarbons in absence of inhibitors and accelerators • Oxidation of simple hydrocarbons in presence of antioxidants • Oxidation of simple hydrocarbons in presence of metal catalysts • Weak chemiluminescence during hydrocarbon autoxidation • Qualitative aspects of autoxidation of saturated polyolefins • Quantitative aspects of autoxidation of saturated polyolefins • Investigation of polyolefin oxidation by various techniques.

REMBAUM and SHEN Biomedical Polymers

edited by ALAN REMBAUM, *California Institute of Technology, Pasadena* and MITCHEL SHEN, *University of California, Berkeley*

304 pages, illustrated. 1971

A collection of papers given at the Symposium on Biomedical Polymers held in Pasadena in 1969. Of interest to scientists in polymer chemistry, polymer physics, biochemistry, bioengineering, materials science, pharmacology, and surgery.

CONTENTS: Problems in blood-tissue reactions to polymeric materials, *B. Zweifach*. Past, present, and future of artificial kidney treatment, *B. Barbour*. The chemistry and properties of the medical-grade silicones, *S. Braley*. Correlation of the surface charge characteristics of polymers with their antithrombogenic characteristics, *S. Srinivasan and P. Sawyer*. Persistent polarization in polymers and blood compatibility, *P. Murphy, A. Lacroix, S. Merchant, and W. Bernhard*. Selection, characterization, and biodegradation of surgical epoxies, *A. Cupples and R. Schubert*. Foreign body reactions to plastic implants, *D. Ocumpaugh and H. Lee*. Rapid in vitro screening of polymers for bio-compatibility, *C. Homsy, K. Ansevin, W. O'Bannon, S. Thompson, R. Hodge, and M. Estrella*. Improved membranes for hemodialysis, *F. Martin, H. Shuey, and C. Saltonstall, Jr*. Control of polymer morphology for biomedical applications, 1. Hydrophilic polycarbonate membranes for dialysis, *R. Kesting*. Surgical adhesives in ophthalmology, *M. Refojo*. Medical uses for polyelectrolyte complexes, *M. Vogel, R. Cross, H. Bixler, and R. Guzman*. Potentialities of a new class of anticlotting and antihemorrhagic polymers, *T. Yen, M. Daver, and A. Rembaum*. Synthesis and properties of a new class of potential biomedical polymers, *A. Rembaum, S. Yen, R. Landel, and M. Shen*. Recognition polymers, *D. Bradley*. The challenge for high polymers in medicine, surgery, and artificial internal organs, *H. Lee and K. Neville*.

RICHARDSON   Optical Microscopy for the Materials Sciences

(Monographs and Textbooks in Material Science Series, Volume 3)

by JAMES H. RICHARDSON, *Aerospace Corporation, El Segundo, California*

(continued)

material science

RICHARDSON *(continued)*
704 pages, illustrated. 1971

Provides in one volume a comprehensive survey of the techniques for preparation and optical examination of specimens in the broad area of materials sciences. A highly useful text for the university or vocational school student studying the microstructure of materials or metallography and also a valuable practical reference for all researchers using microscopy, as well as for engineers and industrial metallographers.

CONTENTS: The Brightfield optical microscope • The microscopy of phase structures • Photomicrography • Photomacrography • Specimen preparation • Reagents and techniques for specimen preparation • Laboratory safety • Examination of the specimen • Accessories • Laboratory design.

ROGERS *Permselective Membranes*

edited by CHARLES E. ROGERS, *Case Western Reserve University, Cleveland, Ohio*

224 pages, illustrated. 1971

Encompasses a broad range of topics pertaining to the expanding field of permselective membranes, including theoretical aspects of transport behavior, new and unusual methods for the preparation or modification of membrane materials, and the effects of experimental conditions on the permselectivity of membranes to both ionic and nonionic penetrants. Of value to biophysicists, polymer chemists, physicists, biochemists, and chemical engineers, as well as biologists and physiologists.

CONTENTS: Transport of dissolved oxygen through silicone rubber membrane, *S. Hwang, T. Tang, and K. Kammermeyer*. Gas transport in segmented block copolymers, *K. Ziegel*. Transport of noble gases in poly(methyl acrylate), *W. Burgess, H. Hopfenberg, and V. Stannett*. Permeation of gases at high pressures, *S. Stern, S. Fang, and R. Jobbins*. Permeation of gases through modified polymer films III. Gas permeability and separation characteristics of gamma-irradiated Teflon FEP copolymer films, *R. Huang and P. Kanitz*. Theoretical interpretation of the effect of mixture composition on separation of liquids in polymers, *M. Fels and R. Huang*. Permselectivity of solutes in homogeneous water-swollen polymer membranes, *H. Yasuda and C. Lamaze*. Ion-exchange selectivity coefficients in the exchange of calcium, strontium, cobalt, nickel, zinc, and cadmium ions with hydrogen ion in variously cross-linked polystyrene sulfonate cation exchangers at 25°C, *M. Reddy and J. Marinsky*. Ion and water transport through permselective membranes, *N. Lakshminarayanaiah*. Permeability of cellulose acetate membranes to selected solutes, *H. Lonsdale, B. Cross, F. Graber, and C. Milstead*. Transport through permselective membranes, *C. Rogers and S. Sternberg*.

SCHEY *Metal Deformation Processes: Friction and Lubrication*
(Monographs and Textbooks in Material Science Series, Volume 1)

edited by JOHN A. SCHEY, *University of Illinois at Chicago Circle*

822 pages, illustrated. 1970

A comprehensive treatment of all aspects of friction and lubrication in metal deformation processes. Of aid to metallurgists, mechanical engineers, chemists, and physicists.

CONTENTS: Background and system of approach, *J. Schey*. Friction effects in metalworking processes, *J. Schey*. Friction, lubrication, and wear mechanisms, *C. Riesz*. Lubricants, *C. Riesz*. Lubricant properties and their measurements, *J. Schey*. Rolling lubrication, *J. Schey*. Wire drawing lubrication, *J. Newnham*. Hot extrusion lubrication, *S. Kalpakjian*. Forging lubrication, *S. Kalpakjian*. Cold forging and cold extrusion lubrication, *J. Newnham*. Sheet metal working lubrication, *J. Newnham*.

SEGAL *High-Temperature Polymers*

edited by CHARLES L. SEGAL, *North American Aviation, Inc., Canoga Park, California*

208 pages, illustrated. 1967

CONTENTS: Introduction, *C. L. Segal*. Thermally stable polymers with aromatic recurring units, *C. S. Marvel*. Inorganic polymer chemistry, *J. R. Van Wazer*. Kinetics and gaseous products of thermal decomposition of polymers, *H. L. Friedman*. Studies of stability of condensation polymers in oxygen-containing atmospheres, *R. T. Conley*. Thermal degradation of polymers. III: Mass spectrometric thermal analysis of condensation polymers, *G. P. Shulman*. Viscoelastic relaxation mechanism of inorganic polymers. V: Counterion effects in bulk polyelectrolytes, *A. Eisenberg, S. Saito, and T. Sasada*. Thermal stability of carborane-containing polymers, *J. Green and N. Mayes*. Synthesis and thermal stability of structurally related aromatic Schiff bases and acid amides, *A. D. Delman, A. A. Stein, and B. B. Simms*. New high-temperature polymers. II: Ordered aromatic copolyamides containing fused and multiple ring systems, *F. Dobinson and J. Preston*. Synthesis of fusible branched polyphenylenes, *N. Bilow and L. J. Miller*.

SEGAL, SHEN, and KELLEY
Polymers in Space Research

edited by CHARLES L. SEGAL, *Whittaker Corporation, San Diego*, MITCHEL SHEN, *University of California, Berkeley*, and FRANK N. KELLEY, *Air Force Rocket Propulsion Laboratory, Edwards, California*. Associate Editors: GEORGE F. PEZDIRTZ, *NASA Langley Research Center, Hampton, Virginia*, and W. DAVID ENGLISH, *Astropower Laboratory, McDonnell Douglas Aeronautics Company, Huntington Beach, California*

480 pages, illustrated. 1970

CONTENTS:

Part I: Recent Developments in the Synthesis, Characterization, and Evaluation of Thermally Stable Polymers

Introduction, *C. Segal and G. Pezdirtz*. Aromatic polymers: Single- and double-stranded chains, *J. Stille*. Thermally stable spiropolymers, *J. Hodgkin and J. Heller*. Isomeric and substituent effects in some dibenzoylbenzene-diamine polymers, *A. Volpe, L. Kaufman, and R. Dondero*. Arylsulfimide polymers. III. The syntheses of some monomeric aryl-1,2-disulfonic acids and derivatives, *G. D'Alelio, Y. Giza, and D. Feigl*. Properties of heterocyclic condensation polymers, *G. Berry and T. Fox*. Relative thermophysical properties of some polyimidazopyrrolones, *R. Jewell*. Thermal decomposition of polyimides in vacuum, *T. Johnston and C. Gaulin*. Thermomechanical behavior of an aromatic polysulfone, *J. Gillham, G. Pezdirtz, and L. Epps*. Panel discussion on thermally stable polymers, *C. Segal, J. Stille, G. Pezdirtz, G. D'Alelio, H. Levine, and W. Gibbs*.

Part II: Properties of Polymers at Low Temperatures

Introduction, *M. Shen and W. English*. Relaxation behavior of polymers at low temperatures, *J. Sauer and R. Saba*. Thermal properties of polymers at low temperatures, *W. Reese*. Multiple transitions in polyvinyl alkyl ethers at low temperatures, *W. Schell, R. Simha, and J. Aklonis*. Internal friction study of diluent effect on polymers at cryogenic temperatures, *M. Shen, J. Strong, and H. Schlein*. Stress-strain behavior of adhesives in a lap joint configuration at ambient and cryogenic temperatures, *G. Tiezzi and H. Doyle*. Some properties of nitroso rubbers in fluorine at ambient and cryogenic temperatures, *S. Toy, W. English, W. Crane, and M. Toy*. Cryogenic properties of a polyurethane adhesive, *R. Robbins*. Some effects of structure on a polymer's performance as a cryogenic adhesive, *R. Gosnall and H. Levine*.

Part III: Solid Propellants

Introduction, *F. Kelley*. Recent developments in solid-propellant binders, *H. Marsh, Jr.* Saturated hydrocarbon polymers for solid rocket propellants, *A. Di Milo and D. Johnson*. Preparation and curing of poly (perfluoroalkylene oxides), *J. Zollinger, J. Throckmorton, S. Ting, R. Mitsch, and D. Elrick*. Functionality and functionality distribution measurements of binder prepolymers, *A. Muenker and B. Hudson, Jr.*

SERAFINI and KOENIG
Cryogenic Properties of Polymers

edited by TITO T. SERAFINI, *NASA-Lewis Research Center, Cleveland*, and JACK L. KOENIG, *Case Western Reserve University, Cleveland, Ohio*

312 pages, illustrated. 1968

CONTENTS: Cryogenic positive expulsion bladders, *R. F. Lark*. Adhesives for cryogenic applications, *L. M. Roseland*. Glass-, boron-, and graphite-filament-wound resin composites and liners for cryogenic pressure vessels, *M. P. Hanson*. Mechanical behavior of poly(ethylene terephthalate), *I. M. Ward*. Effect of film processing on cryogenic properties of poly(ethylene terephthalate), *R. E. Eckert and T. T. Serafini*. Mechanical properties of epoxy resins and glass/epoxy composites at cryogenic temperatures, *L. M. Soffer and R. Molho*. Mechanical relaxation of poly-4-methyl-pentene-1 at cryogenic temperatures, *M. Takayanagi and N. Kawasaki*. Transitions in glasses at low temperatures, *R. A. Haldon, W. J. Schell, and R. Simha*. Mechanical behavior of poly(ethylene terephthalate) at cryogenic temperatures, *C. D. Armeniades, I. Kuriyama, J. M. Roe, and E. Baer*. Infrared studies of chain folding in polymers II. Poly(ethylene terephthalate), *J. L. Koenig and M. J. Hannon*. Crystallization of poly(ethylene terephthalate) from the glassy amorphous state, *G. S. Y. Yeh and P. H. Geil*. Strain-induced crystallization of poly(ethylene terephthalate), *G. S. Y. Yeh and P. H. Geil*. Molecular motion in polytetrafluoroethylene at cryogenic temperatures, *E. S. Clark*. Synthesis of ultrahigh molecular weight poly(ethylene terephthalate), *L.-C. Hsu*. Development of vulcanizable elastomers suitable for use in contact with liquid oxygen, *P. D. Schuman, E. C. Stump, and G. Westmoreland*. Synthesis of fluorinated polyurethanes, *R. Gosnell and J. Hollander*.

SKEIST Reviews in Polymer Technology

edited by IRVING SKEIST, *Skeist Laboratories, Inc., Livingston, New Jersey*

260 pages, illustrated. 1972

Consists of intensive, up-to-date reviews on various aspects of polymer technology. Directed to chemists, engineers, technicians, commercial planners, and others working in the polymers and plastics fields.

CONTENTS: Coupling agents as adhesion promoters, *P. Cassidy and W. Yager*. Processing powdered polyethylene, *A. Zimmerman*. Plastics and other polymers in building, *I. Skeist and J. Miron*. Fire retardance of polymeric ma-

(continued)

material science

SKEIST (continued)

terials, *I. Einhorn.* Recent advances in photo-cross-linkable polymers, *G. Delzenne.* Organic colorants for polymers, *T. Reeve.*

SOLOMON Step-Growth Polymerizations
(Kinetics and Mechanisms of Polymerization Series, Volume 3)

edited by DAVID H. SOLOMON, *C.S.I.R.O., Melbourne, Australia*

416 pages, illustrated. 1972

Presents a critical and constructive assessment of developments in step-growth polymerization and considers the application of theoretical concepts to commercial systems. Highly recommended to students of polymer science and researchers working in the area of step-growth polymerization, including polymer chemists and other scientists in the paint, coatings, and plastics industries.

CONTENTS: Polyesterification, *D. H. Solomon.* Polyamides, *D. C. Jones and T. R. White.* Polyurethanes: The chemistry of the diisocyanate-diol reaction, *D. J. Lyman.* Cyclopolycondensation, *P. M. Hergenrother.* The reactions of formaldehyde with phenols, melamine, aniline, and urea, *M. F. Drum and J. R. LeBlanc.* Diels-Alder polymerization, *W. J. Bailey.* Inorganic polymers, *J. R. MacCallum.*

STEWART Infrared Spectroscopy: Experimental Methods and Techniques

by JAMES E. STEWART, *Durrum Instrument Corporation, Palo Alto, California*

656 pages, illustrated. 1970

A guide to instrumentation and experimental methods and techniques for the infrared spectroscopist. Primarily for those involved with spectroscopy research and for graduate students in chemistry and physics interested in spectroscopy.

CONTENTS: Infrared spectroscopy • The infrared spectrophotometer • Elements of geometric optics • Elements of physical optics • Optical components of infrared spectrophotometers • Optical systems of infrared spectrophotometers • Slit functions and spectral modulation transfer functions of monochromators • Interference spectroscopy • Mechanics of infrared spectrophotometers • Elements of electronics • Infrared detectors • Electronic systems of infrared spectrophotometers • Electromechanical transfer functions of infrared spectrophotometers • Photometric accuracy of infrared spectrophotometers • Experimental methods of infrared transmission spectroscopy • Experimental methods of infrared reflection spectroscopy • Experimental methods of infrared emission spectroscopy • Advanced methods of infrared spectroscopy.

SZEKELY Blast Furnace Technology: Science and Practice

edited by JULIAN SZEKELY, *State University of New York, Buffalo*

414 pages, illustrated. 1972

Represents the efforts of academic and industrial researchers, metallurgists, plant operators, and designers concerned with the most up-to-date aspects of ironmaking technology. Valuable reading for all production engineers, plant operators, and designers concerned with ironmaking technology, and also of importance to metallurgical engineers, research scientists—chemists, physicists, engineers—and students in this field.

CONTENTS: Single particle studies applied to direct reduction and blast furnace operations, *R. Bleifuss.* Structural effects in gas–solid reactions, *J. Szekely and J. Evans.* The use of catalysts to enhance the rate of Boudouard's reaction in direct reduction metallurgical processes, *Y. Rao and B. Jalan.* Thirty psi high top–gas pressure operation at NSC Nagoya works, *T. Yatsuzuka, Y. Yamada, and A. Tayama.* Practical application of mathematical models in ironmaking, *D. Christie, C. Kearton, and R. Thomas.* A mass-transport model of erosion of the carbon hearth of the iron blast furnace, *J. Elliott and J. Popper.* Contribution to the study of the reaction mechanism occurring in high temperature zone of the blast furnace, *R. Vidal and A. Poos.* The place of direct reduction in a modern blast furnace–BOF plant, *J. Peart and D. George.* The blast furnace control problem, *J. A. Laslo.* Modern blast furnace design in Germany, *F. Lenger.* The blast furnace–a transition, *F. Berczynski.* Projected performance of a blast furnace with prereduced burdens, *J. Agarwal.*

SZEKELY The Steel Industry and the Environment

edited by JULIAN SZEKELY, *Center for Process Metallurgy, State University of New York, Buffalo*

312 pages, illustrated. 1973

Contains the proceedings of the Second C. C. Furnas Memorial Conference on The Steel Industry and The Environment held at the State University of New York at Buffalo in November, 1971. Brings together authors representing different viewpoints on the questions raised in examining the interaction of the steel industry and the environment. Of utmost importance to plant operators, designers, metallurgists, environ-

material science

mental scientists, and all others concerned with the impact of the steel industry on the environment.

CONTENTS: The role of the federal government in environmental pollution control, *K. Johnson.* Control of air pollution in the British iron and steel industry, *F. Ireland.* Health and the steel industry environment, *K. Spring.* The economic impact of the installation and operation of pollution abatement devices, *J. Barker.* Desulfurization of coke oven gas: Technology, economics, and regulatory activity, *R. Dunlap, W. Gorr, and M. Massey.* Treatment of coldmill wastewaters by ultrahigh-rate filtration, *C. Symons.* Experience with pollution abatement, *C. Black and W. Sebesta.* The interaction of the socioeconomic and ecological environment in American steel mill towns, *L. Thaxton and R. Genton.* Plant availability versus clean air: An economic dilemma that can be solved, *R. Heller.* A survey of wastewater treatment techniques for steel mill effluents, *T. Centi.* Emission of sulfurous gases from blast-furnace slags, *R. Kaplan and G. Ringstorff.* Treatment of waste gases from the basic oxygen furnace in West Germany, *E. Weber.* On the oxidation of cyanides in the stack region of the blast furnace, *H. Sohn and J. Szekely.* Reclaimed scrap and solid metallics for steelmaking, *J. Elliott.*

TALLAN Electrical Conductivity in Ceramics and Glass

(Ceramics and Glass: Science and Technology Series, Volume 4)

edited by NORMAN M. TALLAN, *Aerospace Research Laboratories, Wright-Patterson Air Force Base, Ohio*

in preparation. 1973

A text which thoroughly describes several aspects of the electrical conductivity of ceramic materials, and discusses their conductivity, physical dependence on their electronic and ionic defect structures, and the transport mechanisms by which charge and mass move through ceramic materials. Additionally stressed is the use of conductivity measurements to characterize the defect structure and transport properties of ceramics. A great aid to advanced students in materials science, ceramics, and glass.

CONTENTS: General concepts of electrical transport, *D. Adler.* Experimental techniques, *R. Blumenthal and M. Seitz.* Defect structure of ceramic materials, *R. Brook.* Electronic conduction mechanisms, *I. Bransky and J. Wimmer.* Controlled valency effects in electronic conductors, *J. Wagner.* Highly conducting ceramics and the conductor-insulator transition, *J. Honig and R. Vest.* Ionic conductivity and electrochemistry of crystalline ceramics, *J. Patterson.* Conductivity of glass and other amorphous materials, *J. Mackenzie.* Microstructural and polyphase effects, *J. Wimmer and H. Graham.*

TSURUTA and O'DRISCOLL Structure and Mechanism in Vinyl Polymerization

edited by TEIJI TSURUTA, *Department of Synthetic Chemistry, Faculty of Engineering, University of Tokyo,* and KENNETH F. O'DRISCOLL, *Department of Chemical Engineering, State University of New York, Buffalo*

552 pages, illustrated. 1969

Presents a general survey of studies on this subject in terms of physical organic chemistry. Topics are organized to focus on the most important chemical features of vinyl compounds and their response to variations in chemical and physical circumstances.

CONTENTS: Historical development of the theory of the reactivity of vinyl monomers, *M. Imoto.* Structure and reactivity of vinyl monomers, *T. Tsuruta.* Initiation in free radical polymerization, *K. F. O'Driscoll and P. Ghosh.* Termination mechanism in radical polymerization, *A. M. North and D. Postlethwaite.* Organometallic compounds as radical-type initiators for vinyl polymerization, *S. Inoue.* Heterogeneous metal peroxides, *T. Otsu.* Polymerization of α, β-disubstituted olefins, *Y. Minoura.* Polymerization of α, β-unsaturated carbonyl compounds, *D. M. Wiles.* Cationic polymerization of vinyl monomers by metal alkyl catalysts, *T. Saegusa.* Rate constants of elementary reactions in cationic polymerization, *T. Higashimura.* Elementary steps in anionic vinyl polymerization, *J. Smid.* Molecular rearrangements in polymerization of vinyl monomers, *A. D. Ketley and L. P. Fisher.*

VOGL Polyaldehydes

edited by OTTO VOGL, *Central Research Division, E. I. duPont de Nemours & Company, Wilmington, Delaware*

152 pages, illustrated. 1967

CONTENTS: Preface, *O. Vogl.* Polyaldehydes: introduction and brief history, *O. Vogl.* Polymerization of formaldehyde, *N. Brown.* Polymerization and copolymerization of trioxane, *M. B. Price and F. B. McAndrew.* Polymerization of aliphatic aldehydes, *O. Vogl.* Polymers of haloaldehydes, *I. Rosen.* NMR studies of polyaldehydes, *E. G. Brame, Jr. and O. Vogl.* Polymerization of fluorothiocarbonyl compounds, *W. H. Sharkey.* Crystal structure of polyaldehydes, *P. Corradini.* Morphology of polyoxymethylene, *P. H. Geil.*

VOGL and FURUKAWA Polymerization of Heterocyclics

edited by OTTO VOGL, *Department of Polymer Science and Engineering, Univer-*

(continued)

material science

VOGL and FURUKAWA *(continued)*

sity of Massachusetts, Amherst, and JUNJI FURUKAWA, *Kyoto University, Japan*

216 pages, illustrated. 1973

Reviews the polymerization of cyclic ethers and thio ethers, lactones, and lactams. Also covers preparation, polymerization, and properties of perfluoro epoxides, kinetics of cyclic ether polymerization, and the influence of ring strain on the rate of polymerization and living polymers based on cationic ring opening polymerization. Valuable to scientists interested in polymer science, heterocyclic chemistry, polymer engineering, and materials science.

CONTENTS: Introduction, *J. Furukawa.* Polymerization of cyclic ethers, *T. Saegusa.* Polymerization of perfluoro epoxides, *H. Eleuterio.* Specific nature of the polymerization of heterocyclics, *N. Enikolpoyan.* New trioxane copolymers, *H. Cherdron.* Alkylene sulfide polymerizations, *F. Lautenschlaeger.* Lactone polymerization and polymer properties, *G. Brode and J. Koleske.* Lactam polymerization, *J. Sebenda.*

WALKER and THROWER *Chemistry and Physics of Carbon: A Series of Advances*

a series edited by PHILIP L. WALKER and PETER A. THROWER, *Department of Material Sciences, Pennsylvania State University, University Park*

Vol. 1 400 pages, illustrated. 1965
Vol. 2 400 pages, illustrated. 1966
Vol. 3 464 pages, illustrated. 1968
Vol. 4 416 pages, illustrated. 1968
Vol. 5 400 pages, illustrated. 1969
Vol. 6 368 pages, illustrated. 1970
Vol. 7 424 pages, illustrated. 1970
Vol. 8 480 pages, illustrated. 1973
Vol. 9 272 pages, illustrated. 1973
Vol. 10 288 pages, illustrated. 1973
Vol. 11 in preparation. 1974

CONTENTS:

Volume 1: Dislocations and stacking faults in graphite, *S. Amelinckx, P. Delavignette, and M. Heerschap.* Gaseous mass transport within graphite, *G. F. Hewitt.* Microscopic studies of graphite oxidation, *J. M. Thomas.* Reactions of carbon with carbon dioxide and steam, *S. Ergun and M. Mentser.* Formation of carbon from gases, *H. B. Palmer and C. F. Cullis.* Oxygen chemisorption effects on graphite thermoelectric power, *P. L. Walker, Jr., L. G. Austin, and J. J. Tietjen.*

Volume 2: Electron microscopy of reactivity changes near lattice defects in graphite, *G. R. Hennig.* Porous structure and adsorption properties of active carbons, *M. M. Dubinin.* Radiation damage in graphite, *W. N. Reynolds.* Adsorption from solution by graphite surfaces, *A. C. Zettlemoyer and K. S. Narayan.* Electronic transport in pyrolytic graphite and boron alloys of pyrolytic graphite, *C. A. Klein.* Activated diffusion of gases in molecular-sieve materials, *P. L. Walker, Jr., L. G. Austin and S. P. Nandi.*

Volume 3: Nonbasal dislocations in graphite, *J. M. Thomas and C. Roscoe.* Optical studies of carbon, *S. Ergun.* Action of oxygen and carbon dioxide above 100 millibars on "pure" carbon, *F. M. Lang and P. Magnier.* X-ray studies of carbon, *S. Ergun.* Carbon transport studies for helium-cooled high-temperature nuclear reactors, *M. R. Everett, D. V. Kinsey, and E. Römberg.*

Volume 4: X-ray diffraction studies on carbon and graphite, *W. Ruland.* Vaporization of carbon, *H. B. Palmer and M. Shelef.* Growth of graphite crystals from solution, *S. B. Austerman.* Internal friction studies on graphite, *T. Tsuzuku and M. H. Saito.* Formation of some graphitizing carbons, *J. D. Brooks and G. H. Taylor.* Catalysis of carbon gasification, *P. L. Walker, Jr., M. Shelef, and R. A. Anderson.*

Volume 5: Deposition, structure and properties of pyrolytic carbon, *J. C. Bokros.* The thermal conductivity of graphite, *B. T. Kelly.* The study of defects in graphite by transmission electron microscopy, *P. A. Thrower.* Intercalation isotherms on natural and pyrolytic graphite, *J. G. Hooley.*

Volume 6: Physical adsorption of gases and vapors of graphitized carbon blacks, *N. N. Avgul and A. V. Kiseleyv.* Graphitization of soft carbons, *J. Maire and J. Méring.* Surface complexes on carbons, *B. R. Puri.* Effects of reactor irradiation on the dynamic mechanical behavior of graphites and carbons, *R. E. Taylor and D. E. Kline.*

Volume 7: The kinetics and mechanism of graphitization, *D. B. Fischbach.* The kinetics of graphitization, *A. Pacault.* Electronic properties of doped carbons, *A. M. Marchand.* Positive and negative magnetoresistances in carbons, *P. Delhaes.* The chemistry of the pyrolytic conversion of organic compounds to carbon, *E. Fitzer, K. Mueller and W. Schaefer.*

Volume 8: The electronic properties of graphite, *I. Spain.* Surface properties of carbon fibers, *D. McKee and V. Mimeault.* The behavior of fission products captured in graphite by nuclear recoil, *S. Yajima.*

Volume 9: Carbon fibers from rayon presursors, *R. Bacon.* Control of structure of carbon for use in bioengineering, *J. Bokros, L. LaGrange, and F. Schoen.* Deposition of pyrolytic carbon in porous solids, *W. Kotlensky.*

Volume 10: The thermal properties of graphite, *B. Kelly and R. Taylor.* Lamellar reactions in graphitizable carbons, *M. Robert, M. Oberlin, and J. Mering.* Methods and mechanisms of growth of synthetic diamond, *F. Bundy, H. Strong, and R. Wentorff, Jr.*

† *Volume edited by Philip L. Walker*

material science

Volume 11: Structure and physical properties of carbon fibers, *W. Reynolds*. Highly oriented pyrolytic graphite, *A. Moore*. Evaporated carbon films, *I. McLintock and J. Orr*. Deformation mechanisms in carbons, *G. Jenkins*.

WARD Chemical Modification of Papermaking Fibers

(Fiber Science Series, Volume 4)
by KYLE WARD, JR., *Institute of Paper Chemistry, Appleton, Wisconsin*
256 pages, illustrated. 1973

Bridges the gap between research and industrial applications in the field of chemical modification of papermaking fibers. Deals with the chemical changes which produce new or improved properties in paper products. Of particular importance to researchers and technologists in the paper, textile, and related industries, and students of polymer and organic chemistry.

CONTENTS: Introduction • Esterification • Etherification • Oxidation • Crosslinking • Graft polymerization onto cellulose.

WASLEY Stress Wave Propagation in Solids: An Introduction

(Monographs and Textbooks in Material Science Series, Volume 5)
by RICHARD J. WASLEY, *Department of Chemistry, University of California, Livermore*
344 pages, illustrated. 1973

Provides the fundamentals necessary for the study of the propagation of short duration, high-intensity, nonelastic, mechanical stress disturbances in solids. The first part of the book treats some of the dynamic analyses of elastic solid media which obey Hooke's law. The last section discusses some of the theoretical and experimental aspects of one-dimensional stress waves and shock loading and response. For scientists and engineers desiring further knowledge in the field of stress wave propagation. Also of interest to advanced undergraduate and graduate students of engineering and physics.

CONTENTS: Elasticity: Quasistatic and dynamic response • Wave propagation in extended media • Wave propagation in semi-extended media: Reflection and refraction • Wave propagation in circular cylindrical rods • Selected applications of concepts of elasticity • Nonelastic material behavior • One-dimensional stress wave investigations • Nonelastic (shock) one-dimensional strain wave investigations.

WEINBERG Tools and Techniques in Physical Metallurgy

In 2 Volumes
edited by FRED WEINBERG, *University of British Columbia, Vancouver*
Vol. 1 416 pages, illustrated. 1970
Vol. 2 376 pages, illustrated. 1970

Aids the non-specialist in understanding and making use of the new instruments and techniques of physical metallurgy.

CONTENTS:
Volume 1: Temperature measurement, *R. Bedford, T. Dauphinee, and H. Preston-Thomas*. X-ray diffraction, *C. M. Mitchell*. Crystal growth and alloy preparation, *F. Weinberg and J. T. Jubb*. Quantitative metallography, *J. R. Blank and T. Gladman*. Metallography, *H. E. Knechtel, W. F. Kindle, J. L. McCall, and R. D. Buchheit*.
Volume 2: Electron microscopy, *E. Smith*. Scanning electron microscopy, *O. Schaaber*. Field-ion microscopy, *B. Ralph*. Thermionic-emission microscopy, *W. L. Grube and S. R. Rouze*. Electron-probe microanalysis, *L. C. Brown and H. Thresh*. Emission spectrography and atomic absorption spectrophotometry, *G. L. Mason*.

WILSON Radiation Chemistry of Monomers-Polymers-Plastics

by JOSEPH E. WILSON, *Department of Chemistry, Bishop College, Dallas, Texas*
in preparation. 1974

Provides an up-to-date survey of the radiation chemistry of monomers, polymers, and plastics. Gives essential information on radiation properties, measurement, and detection, and the primary chemical results of the interaction of radiation with matter. Of particular interest to polymer and radiation chemists.

CONTENTS: (tentative): Types and sources of radiation • Fundamental effects of the irradiation of matter • Short-term chemical effects of radiation absorption • Radiation chemistry of small molecules • Radiolytic polymerization in homogeneous systems • Radiolytic polymerization in the solid state • Radiation-induced polymerization in thermosetting, polyester, and emulsion systems • Irradiation of polymers: Crosslinking versus scission • Radiolytic grafting of monomers on polymeric films • Radiolytic grafting on fibers.

YOCUM and NYQUIST Functional Monomers: Their Preparation, Polymerization, and Applications

In 2 Volumes
edited by RONALD H. YOCUM, *The Dow Chemical Company, Freeport, Texas*, and

(continued)

material science

YOCUM and NYQUIST (continued)

EDWIN B. NYQUIST, *The Dow Chemical Company, Midland, Michigan*

Vol. 1   712 pages, illustrated. 1973
Vol. 2   321 pages, illustrated. 1973

A practical reference work which deals with functional monomers. Presents a broad technical background on the preparation and polymerization of individual functional monomers and their applications to various areas of industry. Of special interest to both academic and industrial chemists, particularly those working in the paint, coatings, and textile industry.

CONTENTS:

Volume 1: Acrylamide and other alpha, beta, and unsaturated acids, *D. C. MacWilliams*. Reactive halogenated monomers, *C. F. Raley and R. J. Dólinski*. Hydroxy monomers, *E. B. Nyquist*. Sulfonic acids and sulfonate monomers, *D. A. Kangas*.

Volume 2: Reactive heterocyclic monomers, *D. Tomalia*. Acidic monomers, *L. Luskin*. Basic monomers, vinylpyridines and aminoalkyl (METH) acrylates, *L. Luskin*.

ZIEF   Purification of Inorganic and Organic Materials: *Techniques of Fractional Solidification*

edited by MORRIS ZIEF, *J. T. Baker Chemical Co., Phillipsburg, New Jersey*

340 pages, illustrated. 1969

Of interest to the chemist, chemical engineer, and metallurgist.

CONTENTS: Analysis of ultrapure materials, *C. L. Grant*. Optical-emission spectrochemical analysis—arc, spark, and flame, *C. L. Grant*. Spark-source mass spectrography, *P. R. Kennicott*. Atomic-absorption spectroscopy, *J. W. Robinson*. Infrared spectrophotometry, *K. E. Stine and W. F. Ulrich*. Gas-liquid chromatography, *R. A. Keller*. Differential thermal analysis and differential scanning calorimetry, *E. M. Barrall, II, and J. F. Johnson*. Electrical resistance-ratio measurement, *G. T. Murray*. Reduction of cyclohexane content of benzene under steady flow conditions, *J. D. Henry, Jr., M. D. Danyi, and J. E. Powers*. Purification of aromatic amines, *B. Pouyet*. The freezing staircase method, *C. P. Saylor*. Purification of aluminum, *J. L. Dewey*. Concentration of humic acids in natural waters, *J. Shapiro*. Fractionation of polystyrene, *J. D. Loconti*. Purification and growth of large anthracene crystals, *J. N. Sherwood*. Purification of indium antimonide, *A. R. Murray*. Purification of alkaline iodides (KI, RbI, CsI), *D. Ecklin*. Zone melting of metal chelate systems, *K. Ueno, H. Kobayashi, and H. Kaneko*. Purification of dienes, *R. Kieffer*. Purification of kilogram quantities of an organic compound, *J. C. Maire and M. Delmas*. Rapid purification of organic substances, *M. J. van Essen, P. F. J. van der Most, and W. M. Smit*. Investigation of zone-melting purification of gallium trichloride by a radiotracer method, *W. Kern*. Purification of potassium chloride by radio-frequency heating, *R. Warren*. Purification of a metal by electron-beam heating, *R. E. Reed and J. C. Wilson*. Heating by hollowcathode gas discharge, *W. Class*. Continuous zone refining of benzoic acid, *J. K. Kennedy and G. H. Moates*. Purification of naphthalene in a centrifugal field, *E. L. Anderson*. Zone-melting chromatography of organic mixtures, *H. Plancher, T. E. Cogswell and D. R. Latham*. The concentration of flavors at low temperature, *M. T. Huckle*. Containers for pure substances, *E. C. Kuehner and D. H. Freeman*.

ZIEF and SPEIGHTS   *Ultrapurity: Methods and Techniques*

edited by MORRIS ZIEF, *J. T. Baker Chemical Co., Philipsburg, New Jersey*, and ROBERT M. SPEIGHTS, *American Metal Climax, Inc., Golden, Colorado*

720 pages, illustrated. 1972

Brings together for the first time the four essential and interrelated parameters of ultrapurity: preparation, handling, containment, and analysis. Reflects the continuing progress in the preparation of ultrapure chemicals, the explosive growth in developments pertaining to the handling and containment of these materials, as well as the necessity for complete analysis. Directed to all those working in research, development, or analysis of ultrapure products.

CONTENTS: Purification of alkali halides, *F. Rosenberger*. Purification of organic solvents by frontal-analysis chromatography, *H. Engelhardt*. The preparation of pure sodium and potassium, *R. L. McKisson*. Sublimation of phosphorus pentoxide, *R. D. Mounts*. Purification of proteins by membrane ultrafiltration, *G. J. Fallick*. The purification of p-xylene by partial freezing, *R. R. Gruden and M. Zief*. Purification of isopropylbenzene by preparative gas-liquid chromatography, *J. R. Gruden and M. Zief*. The preparation of ultrapure chemicals by fractional distillation, *H. Plancher and W. E. Haines*. Purification by dry-column chromatography, *F. M. Rabel*. Preparation of ultrapure water, *V. C. Smith*. Preparation and characterization of cholesterol, *I. L. Shapiro*. Contamination problems in trace-element analysis and ultrapurification, *D. E. Robertson*. Airborne contamination, *J. A. Paulhamus*. Glass containers for ultrapure solutions, *P. B. Adams*. Vitreous silica, *G. Hetherington and L. W. Bell*. Ceramics, *C. Garnsworthy*. High-purity chemicals—a challenge to practical analysis, *A. J. Barnard, Jr*. Emission spectroscopy, *E. C. Snooks*. Flame spectrophotometric trace analysis, *D. C. Burrell*. Neutron-activation analysis, *J. J. Kelly*. Visible spectrophotometry, *R. H.*

Weiss. Coulometric titration, *G. W. Higgins.* Information sources for ultrapurification and characterization, *T. E. Connolly.*

ZIEF and WILCOX Fractional Solidification

edited by Morris Zief, *J. T. Baker Chemical Company, Phillipsburg, New Jersey,* and William R. Wilcox, *Aerospace Corporation, Los Angeles*

736 pages, illustrated. 1967

CONTENTS: Introduction, *W. R. Wilcox.* **Part I: Basic Principles:** Phase diagrams, *G. M. Wolten and W. R. Wilcox.* Mass transfer in fractional solidification, *W. R. Wilcox.* Constitutional supercooling and microsegregation, *G. A. Chadwick.* Polyphase solidification, *G. A. Chadwick.* Heat transfer in fractional solidification, *W. R. Wilcox.* **Part II: Laboratory Scale Apparatus:** Laboratory scale apparatus, *E. A. Wynne and M. Zief.* Batch zone melting, *E. A. Wynne.* Progressive freezing, *D. Richman, E. A. Wynne, and F. D. Rosi.* Continuous-zone melting, *J. K. Kennedy and G. H. Moates.* Column crystallization, *R. Albertins, W. C. Gates, and J. E. Powers.* Zone precipitation and allied techniques, *I. A. Eldib.* **Part III: Industrial Scale Equipment:** Proabd refiner, *J. G. D. Molinari.* Newton Chambers' process, *J. G. D. Molinari.* Rotary-drum techniques, *J. C. Chaty.* Phillips fractional-solidification process, *D. L. McKay.* Desalination by freezing, *J. C. Orcutt.* **Part IV: Applications:** Ultrapurification, *P. Jannke, J. K. Kennedy, and G. H. Moates.* Ultrapurity in pharmaceuticals, *P. Jannke.* Ultrapurity in electronic materials, *J. K. Kennedy and G. H. Moates.* Ultrapurity in materials research, *J. K. Kennedy, G. H. Moates, and W. R. Wilcox.* Ultrapurity in crystal growth, *G. H. Moates and J. K. Kennedy.* Bulk purification, *J. D. Loconti.* Analytical applications of fractional solidification, *A. S. Yue.* Materials preparation, *D. Richman and F. D. Rosi.* **Part V: Economics:** Economics of fractional solidification, *J. C. Chaty and W. R. Wilcox.* **Part VI: Appendix:** Introduction, *M. Zief and C. E. Shoemaker.* Survey of inorganic materials, *C. E. Shoemaker and R. L. Smith.* Survey of organic materials, *M. Zief.*

──────── OTHER BOOKS OF INTEREST ────────

CUTLER and DAVIS Detergency: Theory and Test Methods
In 2 Parts

(Surfactant Science Series, Volume 5)

edited by W. G. Cutler, and R. Davis, *Whirlpool Corporation, Benton Harbor, Michigan*

Part 1 464 pages, illustrated. 1972
Part 2 in preparation. 1973

JUNGERMANN Cationic Surfactants

(Surfactant Science Series, Volume 4)

edited by Eric Jungermann, *Armour-Dial, Inc., Chicago*

672 pages, illustrated. 1970

LINFIELD Anionic Surfactants

(Surfactant Science Series)

edited by Warner M. Linfield, *U.S. Department of Agriculture, Philadelphia, Pennsylvania*

in preparation. 1974

MATTSON and MARK Activated Carbon: Surface Chemistry and Adsorption from Solution

by James S. Mattson, *Rosenstiel School of Marine and Atmospheric Sciences, University of Miami, Florida,* and Harry B. Mark, Jr., *Department of Chemistry, University of Cincinnati, Ohio*

248 pages, illustrated. 1971

PATRICK Treatise on Adhesion and Adhesives

edited by Robert L. Patrick, *Alpha Research and Development, Inc., Blue Island, Illinois*

Vol. 1 Theory
496 pages, illustrated. 1967
Vol. 2 Materials
568 pages, illustrated. 1969
Vol. 3 Special Topics
264 pages, illustrated. 1973

SCHICK Nonionic Surfactants

(Surfactant Science Series, Volume 1)

edited by Martin J. Schick, *Central Research Laboratories, Interchemical Corporation, Clifton, New Jersey*

1,120 pages, illustrated. 1967

material science

SHINODA   Solvent Properties of Surfactant Solutions
(Surfactant Science Series, Volume 2)
edited by Kozo Shinoda, *Department of Chemistry, Yokohama National University, Japan*
376 pages, illustrated. 1967

SLADE and JENKINS
Thermal Analysis
(Techniques and Methods of Polymer Evaluation Series, Volume 1)
edited by Philip E. Slade, Jr., *Monsanto Company, Pensacola, Florida*, and Lloyd T. Jenkins, *Chemstrand Research Center, Durham, North Carolina*
264 pages, illustrated. 1966

SLADE and JENKINS
Thermal Characterization Techniques
(Techniques and Methods of Polymer Evaluation Series, Volume 2)
edited by Philip E. Slade, Jr., *Monsanto Company, Pensacola, Florida* and Lloyd T. Jenkins, *Chemstrand Research Center, Durham, North Carolina*
384 pages, illustrated. 1970

STEVENS   Characterization and Analysis of Polymers by Gas Chromatography
(Techniques and Methods of Polymer Evaluation Series, Volume 3)
by Malcolm P. Stevens, *American University of Beirut, Lebanon*
216 pages, illustrated. 1969

SWISHER   Surfactant Biodegradation
(Surfactant Science Series, Volume 3)
by R. D. Swisher, *Monsanto Company, St. Louis, Missouri*
520 pages, illustrated. 1970

WALTON   Radome Engineering Handbook: Design and Principles
(Ceramics and Glass: Science and Technology Series, Volume 1)
edited by Jesse D. Walton, Jr., *Georgia Institute of Technology, Atlanta*
616 pages, illustrated. 1970

JOURNALS OF INTEREST

BIOMATERIALS, MEDICAL DEVICES, AND ARTIFICIAL ORGANS
An International Journal

editor: T. F. Yen, *University of Southern California, Los Angeles*

The aim of this new international journal is to bridge the gap between the theoretical aspects and practical applications of artificial organs and other medical devices, and implantation materials. The basic principles responsible for the success of artificial organs are stressed in order to encourage new research in this field.

4 issues per volume

JOURNAL OF MACROMOLECULAR SCIENCE—Chemistry

editor: George E. Ham, *White Plains, New York*

This international journal provides scientists with a cross-section of the outstanding contributions from laboratories around the world—published four and one-half months from publisher's receipt of last manuscript. The fields covered include anionic, cationic, and free-radical addition polymerization and copolymerization, the manifold forms of condensation polymerization, polymer reactions, molecular weight studies, temperature-dependent properties, rheology, effects of radiation of all forms, polymer degradation, and many others.

8 issues per volume

material science

JOURNAL OF MACROMOLECULAR SCIENCE—*Physics*

editor: PHILLIP H. GEIL, *Case Western Reserve University*

A periodical devoted to the publication of significant fundamental contributions concerning the physics of macromolecular solids and liquids. Papers deal with research in transition mechanisms and structure property relationships, the physics of polymer solutions and melts, glassy and rubbery amorphous solids, and individual polymer molecules and natural polymers, as well as all the areas generally contained in polymer state physics.

4 issues per volume

JOURNAL OF MACROMOLECULAR SCIENCE—*Reviews in Macromolecular Chemistry*

editors: GEORGE B. BUTLER, *University of Florida, Gainesville,* and KENNETH F. O'DRISCOLL, *State University of New York, Buffalo,* and MITCHELL SHEN, *University of California, Berkeley*

Topics in this journal are reviews of certain recent chronological periods, and also have the advantage of reflecting the authors' knowledge, interpretation, and concise summary of the state of knowledge in the given area. Because of the nature of the journal, and the short time between completion of a manuscript and its publication, reviews of this nature more closely approximate a current review than could otherwise be accomplished.

2 issues per volume

POLYMER-PLASTICS TECHNOLOGY AND ENGINEERING

editor: LOUIS NATURMAN, *Stamford, Connecticut*

This journal reflects the increasing importance that polymer applications, processing developments, and mass production of new polymer products will have in the coming years. The emphasis of the articles that comprise the journal will also consider plastics technology and engineering as an important new feature of this publication.

2 issues per volume

Examination On-Approval Policy

Our policy allows instructors to examine a particular book for a period of two months without charge. In the event that the book is definitely adopted for a course as a class text, the instructor may retain the copy as *his desk copy,* provided he advises us of the adoption and the number of students enrolled in the class. If, however, the book will not be used as a class text, the instructor may return it or send us his remittance, less the educational discount.

material science

JOURNAL SUBSCRIPTION INFORMATION

Subscriptions are entered on a calendar year basis. When a subscription is entered, it entitles the subscriber to all issues in the particular volume.

All journal subscriptions are processed after payment has been received. Only prepaid orders will receive service.

Cancellations requested for other than publisher's error will be accepted only prior to the publication of the first issue of each current volume, and will be subject to a handling charge.

Please add **foreign postage** for delivery to all countries outside the U.S. and Canada.

Air mail postage is available upon request.

Indexes to journals are bound into the last issue of each volume with the exception of review journals which are not indexed.

Back volume information and prices are available upon request.

A complimentary copy of any journal is available upon request.

Date_____

MARCEL DEKKER, INC.
95 Madison Avenue
New York, New York 10016

Please send me a complimentary copy of the following journal(s):

_____

_____

_____

Name_____

Position_____

Company_____

Address_____

City_____ State_____ Zip_____